STUDY GUIDE

to accompany

ELECTRONIC DEVICES
Fourth Edition

and

ELECTRONIC DEVICES:
Electron-Flow Version
Second Edition

by Thomas L. Floyd

Prepared by

Osama Maarouf

Prentice Hall
Englewood Cliffs, New Jersey   Columbus, Ohio

Cover photo: Copyright © Superstock
Editors: Dave Garza and Judith Casillo
Developmental Editor: Carol Hinklin Robison
Production Editor: Rex Davidson
Cover Designer: Brian Deep
Production Manager: Patricia A. Tonneman
Marketing Manager: Debbie Yarnell

This book was printed and bound by Victor Graphics. The cover was printed by Victor Graphics.

© 1996 by Prentice-Hall, Inc.
A Simon & Schuster Company
Englewood Cliffs, New Jersey 07632

All rights reserved. No part of this book may be reproduced, in any form or by any means, without permission in writing from the publisher.

Printed in the United States of America

10 9 8 7 6 5 4 3 2 1

ISBN: 0-13-398454-0

Prentice-Hall International (UK) Limited, *London*
Prentice-Hall of Australia Pty. Limited, *Sydney*
Prentice-Hall Canada Inc., *Toronto*
Prentice-Hall Hispanoamericana, S. A., *Mexico*
Prentice-Hall of India Private Limited, *New Delhi*
Prentice-Hall of Japan, Inc., *Tokyo*
Simon & Schuster Asia Pte. Ltd., *Singapore*
Editora Prentice-Hall do Brasil, Ltda., *Rio de Janeiro*

PREFACE

You are probably now starting on the second leg of your electronics career. You have mastered some of the basic electronic concepts, such as Ohm's Law. You will now begin to put them together and start to see the place these fundamentals have in the more complicated yet fascinating field of electronics.

Look around you and start to see the many uses of electronics in TV, radio, calculators, watches, and many more uses. Be curious and attempt to learn about these different devices. Many will be beyond your present understanding, but the exposure to these concepts will cause you to want to learn more. This is why you have purchased this book.

The system approach in electronics makes it possible to analyze complex systems using shortcuts, such as block diagrams. These block diagrams allow you to see at a glance what functions the various parts of a complete circuit perform. Using this system approach will allow you to isolate malfunctions to a particular part of a circuit. This speeds up your troubleshooting procedure. As an example, if you are servicing a TV that has a good picture but no sound, then the best place to look for the problem could be in the audio circuits.

Read each chapter of the text and then refer to this volume and read the chapter reviews. Do the chapter quiz problems. A quick glance at your results may show some weak points. Then reread the related material to strengthen your knowledge.

May the concepts presented here draw you toward further studies in your career in electronics.

ACKNOWLEDGMENTS

My thanks to my wife, Aida, for all her support and TLC that she gave me while updating and improving this study guide.

To Aida, Mohamad, and Mae

With all my love

CONTENTS

PREFACE

REVIEW OF KEY POINTS IN CHAPTER 1

INTRODUCTION TO SEMICONDUCTORS   1

ATOMS AND ATOMIC STRUCTURE   1
SEMICONDUCTORS, CONDUCTORS, AND INSULATORS   1
COVALENT BONDING AND CHARGE CARRIERS   1
CONDUCTION IN SEMICONDUCTORS   2
SEMICONDUCTOR TYPES   2
PN JUNCTION AND BIASING   2
TECHNICAL TIPS   3
CHAPTER 1 QUIZ   5

REVIEW OF KEY POINTS IN CHAPTER 2

DIODE APPLICATIONS   9

PN JUNCTION DIODES   9
HALF-WAVE RECTIFIERS   9
FULL-WAVE RECTIFIERS   10
POWER SUPPLY FILTERS   10
DIODE LIMITING AND CLAMPING CIRCUITS   11
TROUBLESHOOTING POWER SUPPLY CIRCUITS   11
TECHNICAL TIPS   12
CHAPTER 2 QUIZ   13

REVIEW OF KEY POINTS IN CHAPTER 3

SPECIAL-PURPOSE DIODES   19

ZENER DIODES   19
ZENER DIODE APPLICATIONS   19
VARACTOR DIODES   20
OPTICAL DIODES   20
OTHER TYPES OF DIODES   21
TECHNICAL TIPS   21
CHAPTER 3 QUIZ   23

REVIEW OF KEY POINTS IN CHAPTER 4

BIPOLAR JUNCTION TRANSISTORS 27

BASIC TRANSISTOR OPERATION 27
DC TRANSISTOR CIRCUIT ANALYSIS 27
CUTOFF AND SATURATION 27
MAXIMUM TRANSISTOR RATINGS 28
THE TRANSISTOR AS AN AMPLIFIER 28
THE TRANSISTOR AS A SWITCH 28
TECHNICAL TIPS 29
CHAPTER 4 QUIZ 31

REVIEW OF KEY POINTS IN CHAPTER 5

TRANSISTOR BIAS CIRCUITS 35

THE DC OPERATING POINT 35
BASE BIAS 35
EMITTER BIAS 36
VOLTAGE-DIVIDER BIAS 36
COLLECTOR-FEEDBACK BIAS 36
TROUBLESHOOTING 37
TECHNICAL TIPS 37
CHAPTER 5 QUIZ 39

REVIEW OF KEY POINTS IN CHAPTER 6

SMALL-SIGNAL BIPOLAR AMPLIFIERS 45

SMALL-SIGNAL AMPLIFIER OPERATION 45
TRANSISTOR AC EQUIVALENT CIRCUITS 45
COMMON-EMITTER AMPLIFIERS 46
COMMON-COLLECTOR AMPLIFIERS 47
COMMON-BASE AMPLIFIERS 47
MULTISTAGE AMPLIFIERS 47
TROUBLESHOOTING 48
TECHNICAL TIPS 48
CHAPTER 6 QUIZ 51

REVIEW OF KEY POINTS IN CHAPTER 7

POWER AMPLIFIERS 55

CLASS A AMPLIFIERS 55
CLASS B AND AB PUSH-PULL AMPLIFIERS 55
CLASS C AMPLIFIERS 56
TECHNICAL TIPS 56
CHAPTER 7 QUIZ 59

REVIEW OF KEY POINTS IN CHAPTER 8

FIELD-EFFECT TRANSISTORS AND BIASING 63

THE JUNCTION FIELD-EFFECT TRANSISTOR (JFET) 63
JFET CHARACTERISTICS AND PARAMETERS 63
JFET BIASING 64
THE METAL OXIDE SEMICONDUCTOR FET (MOSFET) 64
MOSFET BIASING 65
TECHNICAL TIPS 65
CHAPTER 8 QUIZ 67

REVIEW OF KEY POINTS IN CHAPTER 9

SMALL-SIGNAL FET AMPLIFIERS 71

SMALL-SIGNAL FET AMPLIFIER OPERATION 71
FET AMPLIFICATION 71
COMMON-SOURCE AMPLIFIERS 71
COMMON-DRAIN AMPLIFIERS 72
TECHNICAL TIPS 72
CHAPTER 9 QUIZ 73

REVIEW OF KEY POINTS IN CHAPTER 10

AMPLIFIER FREQUENCY RESPONSE 77

GENERAL CONCEPTS 77
THE DECIBELS 77
GAIN ROLL-OFF 78
TECHNICAL TIPS 79
CHAPTER 10 QUIZ 81

REVIEW OF KEY POINTS IN CHAPTER 11

THYRISTORS AND OTHER DEVICES 85

THE SHOCKLEY DIODE 85
SILICON-CONTROLLED RECTIFIER (SCR) 85
SCR APPLICATIONS 86
THE SILICON-CONTROLLED SWITCH (SCS) 86
THE DIAC AND TRIAC 86
THE UNIJUNCTION TRANSISTOR (UJT) 87
THE PROGRAMMABLE UNIJUNCTION TRANSISTOR (PUT) 87
THE PHOTOTRANSISTOR 88
THE LIGHT-ACTIVATED SCR (LASCR) 88
OPTICAL COUPLERS 88
TECHNICAL TIPS 89
CHAPTER 11 QUIZ 91

REVIEW OF KEY POINTS IN CHAPTER 12

OPERATIONAL AMPLIFIERS 95

INTRODUCTION TO OPERATIONAL AMPLIFIERS 95
THE DIFFERENTIAL AMPLIFIER 95
OP-AMP PARAMETERS 96
OP-AMPS WITH NEGATIVE FEEDBACK 96
BIAS CURRENT AND OFFSET VOLTAGE COMPENSATION 97
TECHNICAL TIPS 97
CHAPTER 12 QUIZ 99

REVIEW OF KEY POINTS IN CHAPTER 13

OP-AMP FREQUENCY RESPONSE, STABILITY, AND COMPENSATION 103

BASIC CONCEPTS AND OPEN-LOOP RESPONSE 103
CLOSED-LOOP RESPONSE 103
POSITIVE FEEDBACK AND STABILITY 104
COMPENSATION 104
TECHNICAL TIPS 104
CHAPTER 13 QUIZ 105

REVIEW OF KEY POINTS IN CHAPTER 14

BASIC OP-AMP CIRCUITS   **109**

    COMPARATORS   109
    SUMMING AMPLIFIERS   110
    THE INTEGRATOR AND DIFFERENTIATOR   110
    TROUBLESHOOTING   111
    TECHNICAL TIPS   111
    CHAPTER 14 QUIZ   113

REVIEW OF KEY POINTS IN CHAPTER 15

MORE OP-AMP CIRCUITS   **117**

    INSTRUMENTATION AMPLIFIERS   117
    ISOLATION AMPLIFIERS   117
    OPERATION TRANSCONDUCTANCE AMPLIFIERS (OTAS)   118
    LOG AND ANTILOG AMPLIFIERS   118
    CONVERTERS AND OTHER OP-AMP CIRCUITS   118
    TECHNICAL TIPS   119
    CHAPTER 15 QUIZ   121

REVIEW OF KEY POINTS IN CHAPTER 16

ACTIVE FILTERS   **125**

    BASIC FILTER RESPONSES   125
    FILTER RESPONSE CHARACTERISTICS   126
    ACTIVE LOW-PASS FILTERS   126
    ACTIVE HIGH-PASS FILTERS   127
    ACTIVE BAND-PASS FILTERS   127
    ACTIVE BAND-STOP FILTERS   127
    TECHNICAL TIPS   127
    CHAPTER 16 QUIZ   129

REVIEW OF KEY POINTS IN CHAPTER 17

OSCILLATORS AND THE PHASE-LOCKED LOOP 135

THE OSCILLATOR 135
OSCILLATOR PRINCIPLES 135
OSCILLATORS WITH RC FEEDBACK CIRCUITS 135
OSCILLATORS WITH LC FEEDBACK CIRCUITS 136
NONSINUSOIDAL OSCILLATORS 137
THE 555 TIMER AS AN OSCILLATOR 138
THE PHASE-LOCKED LOOP 138
TECHNICAL TIPS 138
CHAPTER 17 QUIZ 141

REVIEW OF KEY POINTS IN CHAPTER 18

VOLTAGE REGULATORS 147

VOLTAGE REGULATION 147
BASIC SERIES REGULATORS 147
BASIC SHUNT REGULATORS 148
BASIC SWITCHING REGULATORS 148
IC VOLTAGE REGULATORS 148
TECHNICAL TIPS 149
CHAPTER 18 QUIZ 151

APPENDIX

ANSWERS TO CHAPTER QUIZZES 155

REVIEW OF KEY POINTS IN CHAPTER 1
INTRODUCTION TO SEMICONDUCTORS

ATOMS AND ATOMIC STRUCTURE

- The smallest part of an element that will retain the characteristics of that element is an **atom**.

- Electrons are arranged in layers around the nucleus of an atom. Each layer is called a **shell**.

- The **valence** of an atom is related to the number of electrons in the outer shell of an atom.

- A valence electron can be given a charge of energy which will cause it to leave the outer orbit and become a **free electron**.

- The movement of free electrons in a conductor is called **current**.

- The force that will cause a valence electron to become a free electron is **voltage** or electromotive force.

- **Silicon** and **germanium** are commonly used semiconductor materials.

- Semiconductor materials use silicon far more than germanium, mainly because silicon will stand a higher temperature and is inexpensive.

SEMICONDUCTORS, CONDUCTORS, AND INSULATORS

- A material that has conductive properties in between that of a conductor and an insulator is a **semiconductor**.

- A material that has very good conductive properties is a **conductor**.

- A material that has no conductive properties is an **insulator**.

COVALENT BONDING AND CHARGE CARRIERS

- The interaction process of the valence electrons of two or more atoms is known as **covalent bonding**.

- Majority carriers are **free electrons** for an n-type semiconductor and **holes** for a p-type semiconductor.

- Minority carriers are **holes** for an n-type semiconductor and **free electrons** for a p-type semiconductor.

CONDUCTION IN SEMICONDUCTORS

- A voltage applied across a section of silicon will repel the free electrons of the silicon and cause them to move towards the positive terminal of the source. This movement of free electrons, as stated, is current.

- The movement of a free electron away from an atom's outer shell leaves a hole in that shell. As the electrons move towards the positive terminal of the source, the holes move towards the negative. This movement of holes is called conventional current flow.

SEMICONDUCTOR TYPES

- Impurities are added to semiconductor materials. This is called **doping**.

- Two types of semiconductor materials are produced: **n-type** and **p-type**.

- N-type doping produces semiconductor material with extra free electrons.

- P-type doping produces semiconductor material with fewer free electrons.

PN JUNCTION AND BIASING

- A **pn junction** is a piece of silicon doped so that half of it is p-type and the other half is n-type. The area between the doping is called a pn junction.

- Semiconductor material doped to have a pn junction forms a **semiconductor diode**.

- A pn junction is very useful since it will allow current to flow in only one direction.

- **Bias** voltage is the DC voltage applied in circuits containing pn junctions of all types. If this bias voltage causes a pn junction to conduct, it is called **forward-bias**. If the junction does not conduct, then it is **reverse-biased**.

- A forward-bias condition will cause the junction to conduct if the voltage applied to the p-material is more positive with respect to the n-material.

- The voltage drop across a forward-biased junction is about 0.7 V for **silicon** and about 0.3 V for **germanium**.

- Once a junction is forward-biased and conducting a current, the voltage drop across the junction will change very little with changes in forward current (I_F).

- During reverse-bias conditions, the junction acts like an open and drops the full voltage applied to the circuit.

- **Reverse leakage current** is a small current that will flow across a reverse-biased junction.

- **Reverse breakdown** will occur if the reverse potential is too great, causing too much current to flow in the reverse direction through the diode. This will normally damage the diode.

TECHNICAL TIPS

- A typical small diode can be recognized by the number prefix written on it. If the number starts with **1N**, then the device is a diode. A complete number might be **1N4004**. Typical small diodes have a shape similar to resistors, round and with leads from each end. Small diodes are also made out of glass, and these types will break if subjected to a bending force.

- The two terminals of a diode are called the **anode (A)** and the **cathode (K)**. The anode is the positive end; that is, it is the end made of p-material. The opposite end is the n-material. Small diodes are marked with a black band denoting the cathode, or negative end. The anode end is usually unmarked.

- The schematic symbol of a diode indicates the cathode or negative end by a bar as shown in the figure. When forward-biased, the electron current flow is opposite to the direction of the symbol arrow.

- A diode can be tested with an ohmmeter. Connect the positive lead of the meter to the anode of the diode and the negative lead of the meter to the cathode or black band end of the diode. If the diode is good, a low value of resistance should be indicated. Reverse the leads and a high value of resistance should be indicated. These two tests will indicate that the diode is good. If both readings are either high or low, then the diode is faulty.

CHAPTER 1 QUIZ

Student Name _____ Date _____

1. An atom is the smallest particle of an element that retains the characteristics of that element.
   a. true
   b. false

2. A pn structure is called a diode.
   a. true
   b. false

3. Semiconductor material of the n-type has very few free electrons.
   a. true
   b. false

4. The silicon material used in semiconductors is extremely pure with no additives.
   a. true
   b. false

5. A reverse-bias silicon diode has about 0.7 V across it.
   a. true
   b. false

6. The term **bias** in electronics usually means
   a. the value of ac voltage in the signal.
   b. the condition of current through a pn junction.
   c. the value of DC voltages for the device to operate properly.
   d. the status of the diode.

7. Doping of a semiconductor material means
   a. that a glue-type substance is added to hold the material together.
   b. that impurities are added to increase the resistance of the material.
   c. that impurities are added to decrease the resistance of the material.
   d. that all impurities are removed to get pure silicon.

8. The forward voltage across a conducting silicon diode is about
   a. 0.3 V.
   b. 1.7 V.
   c. −0.7 V.
   d. 0.7 V.

9. You have an unknown type of diode in a circuit. You measure the voltage across it and find it to be 0.3 V. The diode might be
    a. a silicon diode.
    b. a germanium diode.
    c. a forward-biased silicon diode.
    d. a transistor.

10. A reverse-biased diode has the _____ connected to the positive side of the source, and the _____ connected towards the negative side of the source.
    a. cathode, anode
    b. cathode, base
    c. base, anode
    d. anode, cathode

11. The movement of free electrons in a conductor is called
    a. voltage.
    b. current.
    c. resistance.
    d. capacitance.

12. A silicon diode is forward-biased. You measure the voltage to ground from the anode at _____, and the voltage from the cathode to ground at _____.
    a. 0 V, 0.3 V
    b. 2.3 V, 1.6 V
    c. 1.6 V, 2.3 V
    d. 0.3 V, 0 V

13. A small amount of current flows across the barrier of a reverse-biased diode. This current is called
    a. forward-bias current.
    b. reverse breakdown current.
    c. conventional current.
    d. reverse leakage current.

14. The boundary between p-type material and n-type material is called
    a. a diode.
    b. a reverse-biased diode.
    c. a pn junction.
    d. a forward-biased diode.

15. Reverse breakdown is a condition when a diode
    a. is subjected to a large reverse voltage.
    b. is reverse-biased and a small leakage current flows.
    c. has no current flowing at all.
    d. is heated up by large amounts of current flowing in the forward direction.

16. As the forward current through a silicon diode increases, the voltage across the diode
    a. increases to a 0.7 V maximum.
    b. decreases.
    c. is relatively constant.
    d. decreases and then increases.

17. A rectifier diode is used to
    a. amplify a signal.
    b. change ac to pulsating DC.
    c. regulate a voltage.
    d. change a DC voltage to an ac voltage.

18. Which statement best describes an insulator?
    a. A material with many free electrons.
    b. A material doped to have some free electrons.
    c. A material with few free electrons.
    d. No description fits.

19. As the forward current through a silicon diode increases, the internal resistance
    a. increases.
    b. decreases.
    c. remains the same.

20. A silicon diode measures a low value of resistance with the meter leads in both positions. The trouble, if any, is
    a. the diode is open.
    b. the diode is shorted to ground.
    c. the diode is internally shorted.
    d. the diode is working correctly.

REVIEW OF KEY POINTS IN CHAPTER 2

DIODE APPLICATIONS

PN JUNCTION DIODES

- Rectifier diodes come under a general classification of diodes called **general-purpose diodes**. These general purpose diodes can be used for many different applications.

- Current flows through a diode when the anode is positive with respect to the cathode. This **forward voltage** is about 0.7 V for a silicon diode and around 0.3 V for a germanium diode.

- Diodes can be tested with an ohmmeter. With the leads in the forward-bias position, the ohmmeter reading should be low. Reversing the leads will produce a high reading.

- A **forward-biased** diode, that is one with current flowing, acts like a closed switch.

- A diode that is **reverse-biased** acts like an open switch.

- A diode's main use is its ability to conduct current in only one direction.

HALF-WAVE RECTIFIERS

- If a diode is placed in a circuit with an ac source, the positive alternation will cause the diode to conduct and the negative alternation will turn off the diode. The result is a series of half-waves or pulsating DC current. This circuit is called a **half-wave rectifier**.

- The average DC value of the output of a half-wave rectifier is found by $V_{AVG} = V_p/\pi$.

- The above formula neglects the effect of the forward voltage across a forward-biased diode. The actual value for the peak output voltage for the above formula is $V_{p(out)} = V_{p(in)} - 0.7$ V. This is for a silicon diode.

- The **peak inverse voltage (PIV)** of a rectifier diode is the maximum negative voltage that it must withstand in reverse bias.

- Each diode has a PIV rating that must not be exceeded, or damage to the diode could result.

- Half-wave rectifiers can obtain their ac voltage from transformers. The peak secondary transformer voltage can be found by applying the turns-ratio formula, $V_s = (N_2/N_1)/V_p$.

- The output frequency of a half-wave rectifier is the same as the input frequency.

- The single diode in a half-wave rectifier is forward-biased and conducts for 180° of the input cycle.

FULL-WAVE RECTIFIERS

- A center-tapped **full-wave rectifier** uses two diodes connected to a center-tapped transformer secondary. One diode conducts on the negative alternation and the other conducts on the positive.

- The **average value** for a center-tapped full-wave rectifier is twice the half-wave value. $V_{avg} = 2V_p/\pi$.

- A center-tapped full-wave rectifier diode must be selected to have a PIV greater than the secondary peak voltage.

- A full-wave **bridge rectifier** uses four diodes in a bridge circuit. This type of rectifier does **NOT** require a center-tapped transformer.

- The output voltage of a bridge rectifier is equal to the transformer secondary voltage, if we neglect the diode drops.

- The PIV rating of bridge rectifier diodes is half that of a center-tapped rectifier circuit.

- The output frequency of either type full-wave rectifiers is twice that of the input frequency.

POWER SUPPLY FILTERS

- A filter is required in a power supply to smooth the pulsating DC output.

- A **capacitor-input filter** utilizes an electrolytic capacitor across the rectifier output.

- The capacitor will produce an output voltage that will have a small ac component called **ripple voltage**.

- The ripple frequency equals the line frequency for a half-wave rectifier.

- The ripple frequency is twice the line frequency for either full-wave rectifier.

- The **ripple factor** (r) indicates the ability of a filter to smooth the DC. $r = V_r/V_{DC}$, where V_r is the rms ripple voltage and V_{DC} is the average DC output voltage.

- A filter capacitor should be as large as necessary to keep the ripple at a low value. If a very large filter capacitor is used, the diodes must be able to withstand the large initial surge current.

- An **inductor input filter** uses an inductor in series with the line current to form an LC filter. This filter is very effective in reducing ripple.

DIODE LIMITING AND CLAMPING CIRCUITS

- A diode can be used to limit or clip part of an ac signal. The diode is often placed across the load. When the diode is reverse-biased, the input appears across the diode and the load. When the signal voltage forward-biases the diode, the output is the forward diode voltage, 0.7 V.

- **Diode limiting** circuits are often called **diode clipping** circuits.

- The clipping or limiting voltage can be adjusted by adding a DC voltage in series with the diode. This DC voltage sets the limiting level. Again the actual limiting voltage is affected by the forward diode voltage.

- A **diode clamper circuit** is also known as a DC restorer. With this circuit the entire ac waveform is shifted positive or negative, adding a DC level to the signal.

- Clamper circuits can also be connected to provide a **voltage doubling** circuit. This type of circuit will provide a DC output voltage equal to twice the peak input voltage. Triple or quadruple voltages can also be achieved.

TROUBLESHOOTING POWER SUPPLY CIRCUITS

- An **open diode** in a half-wave rectifier would be indicated by a 0 V DC output.

- An **open diode** in a full-wave center-tapped rectifier would be indicated by a decrease of output DC voltage, a decrease of ripple frequency to 60 Hz, and an increase in ripple voltage.

- An **open diode** in a bridge rectifier would produce the same symptoms as in a full-wave center-tapped rectifier.

- A **shorted diode** in a bridge rectifier would produce a half-wave voltage at the output. This shorted diode would probably damage the other diode in the pair and the output would be 0 V.

- A **shorted filter capacitor** would produce excessive current through the rectifier diodes and cause them to open. The output voltage would be 0 V.

- A **leaky filter capacitor** will produce a lower than normal DC output voltage and an increase in ripple voltage.

TECHNICAL TIPS

- Diodes are essentially solid-state semiconductor switches. Think of a forward-biased diode as completing a circuit, allowing current to flow through the diode to a load. This is the same action as a switch. If you apply reverse voltages to a diode, then the internal resistance is very high (reverse-bias). This prevents current flow. This is also switch action. Diodes are used extensively in electronics to allow or prevent current flow.

- A tip to remember when servicing a power supply is to place your oscilloscope leads across the output of the power supply in the **ac** mode. Increase the oscilloscope vertical gain until the ripple voltage can be seen. You can then measure the peak-to-peak ripple voltage and also observe the ripple voltage frequency. If the ripple frequency is 60 Hz, you are seeing the output from a half-wave rectifier action. A ripple frequency of 120 Hz indicates full-wave rectification.

- An easy way of remembering the diode placement in a bridge rectifier is as follows. If the diode arrows all point towards the right or **output** of the power supply, then the output will be a **positive DC voltage**. If the arrows point towards the **input** of the supply, then the output voltage will be a **negative DC voltage**. Remember, in drawing a bridge diode arrangement, have all the diodes pointing in the same direction, either towards or away from the output.

- A very large percentage of electronic failures occur in power supply circuits. Careful attention to troubleshooting this type of circuit will be very productive and worthwhile in your electronic career.

- An inductor input filter is commonly called a **choke input** filter. A **choke** is a very common name for an inductor. For example, a radio frequency choke (**RFC**) is just an inductor designed to operate at radio frequencies.

- Here is a shortcut to determining the output of a diode limiting or clipping circuit. If the arrow of the diode symbol points down, the output voltage will be the negative alternation. If the arrow points up, then the output will be the positive alternation. The same tip is useful if the clipping level is changed by a DC supply in series with the diode. A down-pointing arrow will provide an output voltage when the input voltage is negative with respect to the limiting voltage. Of course, the reverse is true if the arrow points up.

- The output of diode clamping circuits or DC restorers can be predicted by the direction of the diode arrows. A clamper circuit does not change the shape of the incoming waveform; it just shifts it in a positive or negative direction. The diode symbol arrow again will indicate the direction of shift. A down arrow will shift the waveform in a negative direction, and an up arrow will shift the waveform in a positive direction.

CHAPTER 2 QUIZ

Student Name _____ Date _____

1. A diode conducts current when reverse-biased and blocks current when forward-biased.
   a. true
   b. false

2. A single diode in a half-wave rectifier conducts for 180° of the input cycle.
   a. true
   b. false

3. The output voltage of a center-tapped full-wave rectifier equals the secondary output voltage.
   a. true
   b. false

4. Ripple voltage is caused by the charging and discharging of the filter capacitor.
   a. true
   b. false

5. Diode limiters add a DC level to an ac signal.
   a. true
   b. false

6. A silicon diode is connected in series with a 5 kΩ resistor and a 5 V battery. If the anode is connected to the positive source terminal, the voltage from the cathode to the negative source terminal is
   a. 0.3 V.
   b. 0.7 V.
   c. 4.7 V.
   d. 4.3 V.

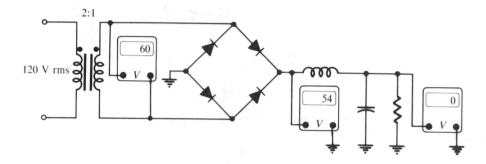

FIGURE 2-1

7. Refer to Figure 2-1. The probable trouble, if any, indicated by these voltages is
   a. one of the diodes is open.
   b. a diode is shorted.
   c. an open transformer secondary.
   d. the filter capacitor is shorted.
   e. no trouble exists.

8. Refer to Figure 2-1. If the voltmeter across the transformer reads 0 V, the probable trouble, if any, would be
   a. one of the diodes is open.
   b. a diode is shorted.
   c. an open transformer secondary.
   d. the filter capacitor is shorted.
   e. no trouble exists.

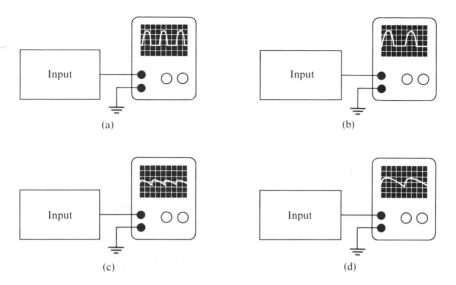

Assume that each scope has the same settings.

FIGURE 2-2

9. Refer to Figure 2-2. Which oscilloscope trace indicates the output from a properly operating half-wave rectifier without a filter?
   a. a
   b. b
   c. c
   d. d

10. Refer to Figure 2-2. Which oscilloscope trace indicates the output from a properly operating full-wave rectifier with a filter?
    a. a
    b. b
    c. c
    d. d

11. Refer to Figure 2-2. Which oscilloscope trace indicates the output from a full-wave rectifier with an open diode?
    a. a
    b. b
    c. c
    d. d

12. Refer to Figure 2-2. The oscilloscope trace in (b) could represent the output from
    a. a full-wave rectifier (no filter) with an open diode.
    b. a normal half-wave power supply.
    c. a filtered full-wave rectifier with an open diode.
    d. a full-wave rectifier with a shorted diode.

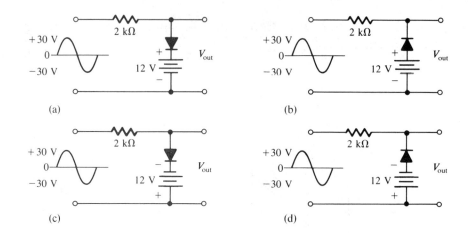

FIGURE 2-3

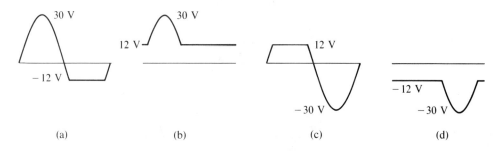

FIGURE 2-4

13. Refer to the circuit in Figure 2-3(a). Refer to the output waveforms shown in Figure 2-4 and select the correct approximate output waveform.
    a. a
    b. b
    × c. c
    d. d

14. Refer to the circuit in Figure 2-3(b). Refer to the output waveforms shown in Figure 2-4 and select the correct approximate output waveform.
    a. a
    — b. b
    c. c
    d. d

15. Refer to the circuit in Figure 2-3(c). Refer to the output waveforms shown in Figure 2-4 and select the correct approximate output waveform.
    a. a
    b. b
    c. c
    × d. d

16

16. Refer to the circuit in Figure 2-3(d). Refer to the output waveforms shown in Figure 2-4 and select the correct approximate output waveform.
    a. a
    b. b
    c. c
    d. d

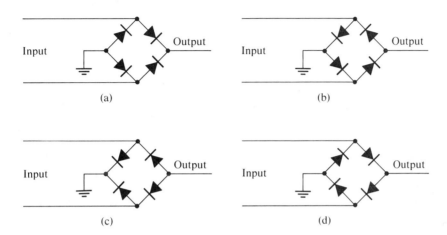

FIGURE 2-5

17. Refer to Figure 2-5. Which diode arrangement will supply a negative output voltage?
    a. a
    b. b
    c. c
    d. d

18. Refer to Figure 2-5. Which diode arrangement will supply a positive output voltage?
    a. a
    b. b
    c. c
    d. d

19. A clamper circuit has an input voltage of 12 V_{p-p}. The diode is pointing down in the schematic diagram. What is the approximate DC output voltage?
    a. 6 V
    b. −6 V
    c. 12 V
    d. −12 V

20. A silicon diode has a voltage to ground of −117 V from the anode. The voltage to ground from the cathode is −117.7 V. The diode is
    a. open.
    b. shorted.
    c. forward-biased.
    d. reverse-biased.

REVIEW OF KEY POINTS IN CHAPTER 3

SPECIAL-PURPOSE DIODES

ZENER DIODES

- A **zener diode** operates in the reverse breakdown region.

- A zener diode is used in the reverse bias position.

- **Avalanche** breakdown is used when the zener voltage is greater than approximately 5 V.

- **Zener** breakdown is used when the zener voltage is less than approximately 5 V.

- If a zener diode is operating in either breakdown mode, the voltage across the zener is essentially constant.

- The zener **resistance** (r_Z) is essentially constant over the operating range of the device.

- The **terminal voltage** (V'_Z) = $V_Z + I_Z r_Z$, where I_Z is the zener current.

ZENER DIODE APPLICATIONS

- A zener diode can be used to regulate a varying DC voltage. This is **line or input regulation**.

- **Load regulation** is a use for a zener diode. In this type of operation, the output voltage to the load is kept constant, within the limits of zener current.

- Load regulation can be determined by calculating the **percent of load regulation**.

- **% Regulation** = $100 \times \dfrac{(V_{NL} - V_{FL})}{V_{FL}}$

  This formula determines the percent of voltage regulation, where V_{NL} is the output voltage with no load and V_{FL} is the output voltage with full load.

- Zener diodes can be used as limiters. The ac input signal will be clipped at the zener voltage and at the zener forward voltage.

VARACTOR DIODES

- A **varactor diode** is used as a **voltage-variable capacitor**.

- In the reverse-bias position, an increase of reverse-bias voltage will increase the capacitance. A decrease of reverse-bias voltage will decrease the capacitance.

- Varactor diodes are used in tuning circuits. If varactors are in a parallel resonant circuit, the resonant frequency can be varied by a varying reverse voltage applied to the varactors.

OPTICAL DIODES

LED　　　　　**PHOTODIODE**

- The **light-emitting diode (LED)** operates in the forward-bias position. When an LED is forward-biased, light is emitted.

- As forward current increases, the amount of emitted light increases.

- The light emitted is directly proportional to the forward current.

- **LED**s come in a variety of colors, such as red, green, or yellow.

- Displays that use **LED**s often are called **seven-segment displays**. These are used for digital clock numbers, for example.

- Another device that conducts with light intensity is the **photodiode**.

- The **photodiode** has a small window to allow light to enter, which activates the pn junction. If the diode is in the reverse-bias position, this incident light will cause current to flow. The more light enters, the more current flows.

- A **photodiode** has a high value of resistance when in the dark. As the light intensity increases, the internal resistance of the photodiode decreases.

- The **photodiode** conducts in one direction only.

OTHER TYPES OF DIODES

SCHOTTKY DIODE   **TUNNEL DIODE**

- A **Schottky diode** is used in high-speed switching circuits.

- A **tunnel diode** has a special characteristic known as **negative resistance**. This diode opposes Ohm's Law during a part of its response curve. That is, as the forward voltage increases, the current decreases.

- A **PIN diode** acts in the reverse-bias position as a constant capacitance, and in the forward position, it acts as a variable resistance.

- The PIN diode is used as a high frequency DC-controlled microwave switch.

- The **step-recovery diode** is a very fast switching diode. This means that this diode can change from on to off very rapidly.

- The **laser diode** normally emits coherent light. An LED emits incoherent light.

- The light from a laser is known as **coherent light**. Coherent light is one that possesses a very narrow band of wavelengths.

- **Laser** is an acronym for **Light Amplification by Simulated Emission of Radiation**.

- The **IMPATT diode** is a fast-switching diode used in microwave oscillators (10–100 GHz).

- The **Gunn diode** is a negative-resistance device used in microwave frequency oscillators.

TECHNICAL TIPS

- Voltage regulation using zener diodes is a very common application. It is important to remember that a series-current-limiting resistor must be used in zener diode applications. This resistor is in the circuit to provide the voltage drop difference between the zener voltage and the source voltage.

- Zener diodes come in a variety of voltage ratings between 1.8 V and 200 V. Zeners are also rated in their power-dissipating capacity or in their current-carrying limits. For example, a zener might be rated at 5.1 V and 1 W. A calculation for the zener current would be $I_Z = 1 \text{ W}/5.1 \text{ V}$. This gives us an I_Z of 196 mA. A zener might be rated in current. In this case the maximum current is known. Do not exceed this maximum zener current.

- Load regulation using zener diodes operates simply in this manner. If the load voltage attempts to decrease due to a change in load resistance, the increase in load current is offset by a decrease in the zener current. The current through the series resistor will not change. The voltage across the load will remain fairly constant. Of course, the opposite changes will occur if the load decreases.

- Line regulation with a zener offers a similar analysis. For example, an increase in line voltage will cause an increase in line current. Since the load voltage will be constant (maintained by the zener), the increased line current will result in an increase in zener current.

- Using the percent regulation formula will result in a percent value. If this number is small, regulation is good. The larger the percent regulation, the more the output voltage will drop with an increase in load current.

- An LED emits light when forward-biased. Though the device is a diode, the forward voltage required to bias it into conduction is greater than the normal 0.7 V we are used to. The forward-bias voltage of some LEDs is around 1.6 V to 2.4 V. A typical current when conducting is about 20 mA. When testing a LED, remember that the forward voltage will be above 0.7 V.

- When using LEDs, always use a series voltage dropping resistor in series with the diode. In performing a lab experiment, for example, you may want to test the LED for proper operation. Do not take the LED and place it across a 5 V power supply. The LED will work, but for a very short time. Use a 330 Ω resistor or, if you want to be safer, use a 1 kΩ resistor in series with the LED.

- Another optoelectronic device worth mentioning is the liquid crystal display (LCD). This device does not emit light, nor is it a light-sensitive device. It is widely used in hundreds of applications requiring alphanumeric displays, from watches to gas station pumps. The LCD consists of a layer of liquid that can pass or block light. This ability is controlled by an electric field which is voltage sensitive. When the liquid is transparent, no field is present. A field will cause the liquid to become opaque. The familiar display of numbers or letters will result. A definite advantage is that the LCD requires very little power. A disadvantage is that the display is visible only in a lighted area.

CHAPTER 3 QUIZ

Student Name _____ Date _____

1. The regulating ability of zener diodes depends upon their ability to operate in a forward-bias condition.
   a. true
   - b. false

2. A photodiode is used in a reverse-bias position, and it will increase conduction as the light intensity increases.
   - a. true
   b. false

3. The larger the percent regulation, the better.
   a. true
   ✗ b. false

4. An LED emits light when forward-biased.
   - a. true
   b. false

5. A tunnel diode has a negative-resistance characteristic.
   - a. true
   b. false

6. An 8.2 V zener has a resistance of 5 Ω. The actual voltage across its terminals when the current is 25 mA is
   a. 8.2 V.
   b. 125 mV.
   - c. 8.325 V.
   d. 8.075 V.

7. A 6.2 V zener is rated at 1 watt. The maximum safe current the zener can carry is
   a. 1.61 A.
   - b. 161 mA.
   c. 16.1 mA.
   d. 1.61 mA.

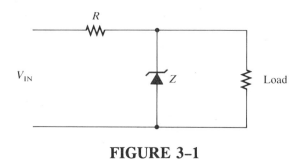

FIGURE 3-1

8. Refer to Figure 3-1. If the load current increases, I_R will _____ and I_Z will _____.
   a. remain the same, increase
   b. decrease, remain the same
   c. increase, remain the same
   d. remain the same, decrease

9. Refer to Figure 3-1. If V_{LOAD} attempts to increase, V_R will
   a. increase.
   b. decrease.
   c. remain the same.

10. Refer to Figure 3-1. If V_{IN} increases, I_Z will
    a. increase.
    b. decrease.
    c. remain the same.

11. Refer to Figure 3-1. If V_{IN} decreases, I_R will
    a. increase.
    b. decrease.
    c. remain the same.

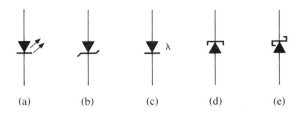

FIGURE 3-2

12. Refer to Figure 3-2. Identify the Schottky diode.
    a. a
    b. b
    c. c
    d. d
    e. e

13. Refer to Figure 3-2. Which symbol is correct for a zener diode?
    a. a
    b. b
    c. c
    d. d
    e. e

14. Refer to Figure 3-2. Find the tunnel diode symbol.
    a. a
    b. b
    c. c
    d. d
    e. e

15. Refer to Figure 3-2. Which symbol is correct for a photodiode?
    a. a
    b. b
    c. c
    d. d
    e. e

16. Refer to Figure 3-2. Which symbol is correct for an LED?
    a. a
    b. b
    c. c
    d. d
    e. e

17. An LED is forward-biased. The diode should be on, but no light is showing. A possible trouble might be
    a. the diode is open.
    b. the series resistor is too small.
    c. none. The diode should be off if forward-biased.
    d. the power supply voltage is too high.

18. The Schottky diode is used
    a. in high-power circuits.
    b. in circuits requiring negative-resistance.
    c. in very fast-switching circuits.
    d. in power supply rectifiers.

19. A tunnel diode is used
    a. in high-power circuits.
    b. in circuits requiring negative-resistance.
    c. in very fast-switching circuits.
    d. in power supply rectifiers.

20. You have an application for a diode to be used in a tuning circuit. A type of diode to use might be
    a. an LED.
    b. a Schottky diode.
    c. a Gunn diode.
    d. a varactor.

REVIEW OF KEY POINTS IN CHAPTER 4

BIPOLAR JUNCTION TRANSISTORS

BASIC TRANSISTOR OPERATION

- A **bipolar junction transistor (BJT)** is a semiconductor device with two pn junctions. In normal operation, the **base-emitter junction (BE)** is forward-biased, while the **base-collector junction (BC)** is reverse-biased.

- When **base current (I_B)** is flowing across the **BE** junction and even though the **BC** junction is reverse-biased, then a current will flow from the emitter to the collector. This current is called the **collector current (I_C)**.

- The current flowing at the emitter is called the **emitter current (I_E)**.

- The relationship of these three currents is $I_E = I_B + I_C$.

DC TRANSISTOR CIRCUIT ANALYSIS

- The ratio of the collector current I_C to the base current I_B is called the **DC current gain** or **DC beta (β_{DC})**. $\beta_{DC} = I_C/I_E$

- Transistor data sheets usually refer to β_{DC} as h_{FE}.

- The ratio of I_C to I_E is called the **DC alpha (α_{DC})**. $\alpha_{DC} = I_C/I_E$

- V_{BE}, when forward-biased, is about 0.7 V.

- Rearranging the equation for β_{DC}, we find that $I_C = \beta_{DC}I_B$. The collector current depends on the flow of the base current. The input current I_B is amplified by the value of β_{DC}.

- The voltage drop across the **collector resistor (R_C)** is $V_{RC} = I_C R_C$.

- The voltage at the collector with respect to ground V_{CE} is found by $V_{CE} = V_{CC} - V_{RC}$.

CUTOFF AND SATURATION

- **Cutoff** is a condition in the operation of a **BJT** when $I_B = 0$ and therefore $I_C = 0$.

- During cutoff, the BE junction is reverse-biased, and only a small amount of **collector leakage current (I_{CEO})** flows.

- The BC junction is also reverse-biased during cutoff.

- **Saturation** is a condition when I_E increases, causing I_C to increase. This increase in I_{CE} forces V_{CE} to decrease to nearly zero. The BC junction becomes forward-biased, and V_{CE} reaches a value called **$V_{CE(sat)}$** (about 0.1 V).

- At saturation the collector current becomes known as **$I_{C(sat)}$**. Further increase in I_B can cause no additional I_C to flow.

- β_{DC} increases directly with an increase in temperature.

MAXIMUM TRANSISTOR RATINGS

- Manufacturers' data sheets list maximum parameters for safe operation of the BJT.

- These maximum ratings usually cover V_{CB}, V_{CE}, V_{BE}, I_C, and **power dissipation $P_{D(max)}$**.

- The product of I_C and V_{CE} must not exceed $P_{D(max)}$. Maximum values of both I_C and V_{CE} must not be used at the same time. Calculate these values with $I_C = P_{D(max)}/V_{CE}$, or rearranging the equation, $V_{CE} = P_{D(max)}/I_C$.

THE TRANSISTOR AS AN AMPLIFIER

- **Amplification** is the process of linearly increasing the amplitude of an electrical signal.

- Since I_B is small compared to I_C, an approximation can be made that $I_C = I_E$.

- Amplification most often deals with ac signals. The **ac emitter current (I_e)** $= V_{in}/r'_e$ where r'_e is the ac resistance of the emitter.

- The ratio of V_{out} to V_{in} is called **voltage gain (A_v)**. $A_v = V_{out}/V_{in}$

- Voltage gain can be calculated by $A_v = R_C/r'_e$.

THE TRANSISTOR AS A SWITCH

- Cutoff, the condition where **$I_B = 0$**, also will cause V_{CE} to equal V_{CC}. In cutoff, then, the BJT is acting like an open switch. No current is flowing.

- Saturation, the condition where I_C is maximum, causes V_{CE} to be near zero. Since there is maximum I_C, the internal resistance between the emitter and the collector is very low. In saturation, the BJT is acting like a closed switch. Maximum current is flowing.

TECHNICAL TIPS

- Cutoff is a very important condition in troubleshooting BJT circuits. A cutoff transistor has the voltage between collector and ground equal to V_{CC}. If you measure this voltage, V_C, and it equals V_{CC}, then there is an open somewhere in the circuit path. This simple clue can give you an excellent start on finding the trouble.

- Another important troubleshooting voltage indication is measuring V_{CE} and finding it is near 0 V. From this, you can see, too much current is flowing. This knowledge, along with other circuit parameters, will give you an advantage in locating the defective component.

- The DC current gain of a BJT, β_{DC}, varies from transistor to transistor. Even devices with the same number may have a β_{DC} that can vary over a three-to-one range. This causes circuit designers to lay out the circuits in various ways to eliminate the effects of this variance. If you replace a defective transistor with another, and the β_{DC} is 200 instead of 100, for example, the new value of I_C could be twice what it should be. The next chapter will outline some circuit changes to handle this problem. Be on the watch for this condition.

- An ohmmeter can be used to check a BJT. Place the leads between the base and emitter. Then reverse the leads. The reading should be high in the reverse-bias position and low in the forward-bias position. Do the same for the BC junction. Last, measure between the emitter and the collector. This reading should be high in both lead positions. If this reading is low, then the EC is shorted.

- When you are connecting transistors, remember that the pin-out diagrams given in the manufacturer's data sheets are bottom views. If you are locating a terminal from above the board, remember to switch the lead around.

CHAPTER 4 QUIZ

Student Name _____ Date _____

1. A BJT consists of three regions: base, emitter, and collector.
   a. true
   b. false

2. β_{DC} is the ratio of I_C to I_E.
   a. true
   b. false

3. BJT transistors have two general types, npn and pnp.
   a. true
   b. false

4. If maximum I_C and maximum V_{CE} are applied to a BJT, maximum power dissipation will not be exceeded.
   a. true
   b. false

5. Both junctions of a BJT should be forward-biased for proper operation.
   a. true
   b. false

6. A BJT has an I_B of 50 µA and a β_{DC} of 75; I_C is:
   a. 375 mA
   b. 37.5 mA
   c. 3.75 mA
   d. 0.375 mA

7. A certain transistor has I_C = 15 mA and I_B = 167 µA; β_{DC} is:
   a. 15
   b. 167
   c. 0.011
   d. 90

8. For normal operation of a pnp BJT, the base must be _____ with respect to the emitter and _____ with respect to the collector.
   a. positive, negative
   b. positive, positive
   c. negative, positive
   d. negative, negative

9. A transistor amplifier has a voltage gain of 100. If the input voltage is 75 mV, the output voltage is:
   a. 1.33 V
   b. 7.5 V
   c. 13.3 V
   d. 15 V

10. A 35 mV signal is applied to the base of a properly biased transistor with an $r'_e = 8\ \Omega$ and $R_C = 1\ k\Omega$. The output signal voltage at the collector is:
    a. 3.5 V
    b. 28.57 V
    c. 4.375 V
    d. 4.375 mV

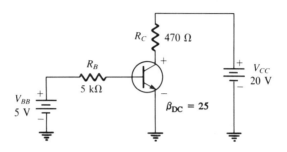

FIGURE 4-1

11. Refer to Figure 4-1; the value of V_{BE} is:
    a. 0.6 V
    b. 0.7 V
    c. 1.2 V
    d. 0.079 V

12. Refer to Figure 4-1; the value of V_{CE} is:
    a. 9.9 V
    b. 9.2 V
    c. 0.7 V
    d. 19.3 V

13. Refer to Figure 4-1; the value of V_{BC} is:
    a. 9.2 V
    b. 9.9 V
    c. −9.9 V
    d. −9.2 V

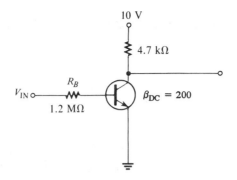

FIGURE 4-2

14. Refer to Figure 4-2; $I_{C(sat)}$ is:
    a. 0.05 mA
    b. 2.13 mA
    c. 1.065 mA
    d. 7.4 mA

15. Refer to Figure 4-2. Determine the minimum value of I_B that will produce saturation.
    a. 0.25 mA
    b. 5.325 µA
    c. 1.065 µA
    d. 10.65 µA

16. Refer to Figure 4-2. Determine the minimum value of V_{IN} that will saturate this transistor.
    a. 13.48 V
    b. 12.08 V
    c. 0.7 V
    d. 9.4 V

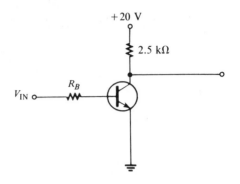

FIGURE 4-3

17. Refer to Figure 4-3. The value of $\beta_{DC} = 100$ and $V_{IN} = 8$ V. Determine $I_{C(sat)}$.
    a. 18 mA
    b. 8 mA
    c. 1.8 mA
    d. 8 µa

18. Refer to Figure 4-3. In this circuit $\beta_{DC} = 100$ and $V_{IN} = 8$ V. The value of R_B that will produce saturation is:
    a. 91.25 kΩ
    b. 9.1 MΩ
    c. 100 kΩ
    d. 150 kΩ

19. Refer to Figure 4-3. The measured voltage, V_{CE}, is 20 V. The transistor is in
    a. saturation.
    b. cutoff.
    c. normal.
    d. not enough data.

20. Refer to Figure 4-3. This circuit comes into your shop for repair. You measure V_{CE} and find it nearly equal to zero. You now know that the transistor is
    a. operating in cutoff.
    b. operating normally.
    c. operating in saturation.
    d. operating below cutoff.

REVIEW OF KEY POINTS IN CHAPTER 5

TRANSISTOR BIAS CIRCUITS

THE DC OPERATING POINT

- Bias voltages are the DC voltages from the BJT elements to ground. These voltages make sure that the BE junction is forward-biased and that the BC junction is reverse-biased.

- Ideally, the DC voltage on the collector (V_C) should be about half of V_{CC}. This will allow the ac portion of the output to swing between the maximum voltage V_{CC} and the minimum voltage $V_C = 0$ V.

- The two maximum operating points for a BJT are cutoff and saturation.

- Characteristic curves for a transistor show the relationships between V_{CE} and I_C for various values of I_B.

- A **load line** is drawn from the point of $I_C = I_{C(sat)}$ to $V_{CE(off)} = V_{CC}$.

- Maximum output in **linear operation** is the operation of the transistor amplifier with the **operating point (Q)** near the center of the load line.

- If the operating point is moved along the load line in either direction towards saturation or cutoff, then the undistorted gain is smaller. Further movement along the load line can cause distortion.

BASE BIAS

- Base bias is sometimes called self-bias or fixed bias.

- In the base-biased circuit, $I_B = \dfrac{V_{CC} - V_{BE}}{R_B}$ and $V_{CE} = V_{CC} - I_C R_C$, where $I_C = \beta_{DC} I_B$

- A disadvantage of this type of bias is that the operating point is dependent on the value of β_{DC}. Temperature and manufacturing variations make the values of β_{DC} change.

EMITTER BIAS

- An emitter-biased circuit uses both a positive and a negative supply voltage.

- In the emitter-biased circuit, the following hold true.

$$I_C \cong \frac{V_{EE} - V_{BE}}{R_E + \frac{R_B}{\beta_{DC}}} \quad \text{and} \quad V_{CE} \cong V_{CC} + V_{EE} - I_C(R_C + R_E)$$

- Realistically, this type of biasing has some disadvantages such that the operating point is dependent on the value of β_{DC} and V_{BE}, both of which change with temperature and current.

VOLTAGE-DIVIDER BIAS

- Voltage divider bias provides good stability of the operating point Q.

- Refer to Figure 5-18 in your textbook, if I_B is much smaller than I_2, then R_1 and R_2 form a simple voltage divider to fix the voltage on the base.

- If I_B is not much larger than I_2, then the **input base resistance** must be used. $R_{IN(base)} = \beta_{DC}R_E$

- If $\beta_{DC}R_E \gg R_2$, then $V_B = [R_2/(R_1 + R_2)]V_{CC}$.

- The emitter current can be found by $I_E = V_E/R_E$.

- Since $I_E \cong I_C$, we find that $V_C \cong V_{CC} - I_C R_C$.

- The voltage divider bias is very stable with the addition of an emitter resistor (R_E). The operating point is independent of the changes that can occur with β_{DC}.

- If pnp transistors are used, it means that the polarities of the bias voltages need to be reversed.

COLLECTOR-FEEDBACK BIAS

- **Collector-feedback bias** provides **negative feedback** to stabilize the position of the Q-point.

- The base bias voltage comes from the collector voltage V_C through a feedback resistor. This feedback is negative because it tends to decrease the value of V_{BE} if β_{DC} increases.

- This type of bias is useful for stabilizing circuits with wide temperature variations.

TROUBLESHOOTING

- In a typical voltage-divider bias circuit, there are only one transistor and four resistors that can fail.

- If R_1 opens, $V_B = 0$ V, $V_E = 0$ V, and $V_C = V_{CC}$.

- If R_2 opens, the transistor will saturate. $V_{CC} = 0$ V and V_B will be about 0.7 V above the saturated value of V_E.

- If R_E opens, the transistor is cutoff, and $V_C = V_{CC}$. V_B is determined by the voltage divider action of R_1 and R_2. V_E will be 0.7 V below V_B.

- If R_C opens, V_B will be low because I_B will be large trying to saturate the transistor. V_E will equal the voltage at V_C. V_C will be about 0.7 more than V_B.

- If the BE junction opens, the voltages will be the same as with R_E open, except that V_E will equal zero.

- If the BC junction opens, the symptoms will be similar to an open R_C, except that $V_C = 0$ V.

TECHNICAL TIPS

- Negative feedback is added to a circuit when an emitter resistor is used. This negative feedback acts to reduce the change to V_{BE} that a signal may produce. Thus, the A_v will be lowered. Of course, this lowering of the voltage gain will also act to stabilize the Q-point so that the operating point will be relatively independent of β. As you can see, a compromise is made between A_v and stability. These compromises are encountered in almost all electronic designs.

- The voltage-divider bias circuits with R_1, R_2, R_C, and R_E are often called H-bias, simply because the circuit looks like an H. All electronic names and symbols are not always uniform throughout the industry. You need to be aware of these variations. Do not be confused by different names and symbols just because they are different from those you have learned.

- The self- or fixed-bias circuit has a disadvantage that can occur under certain conditions. This condition is called thermal runaway. If an increase in temperature increases β_{DC}, then I_C will increase. This change in I_C will heat the transistor more, causing β_{DC} to increase again. This circle can increase until the transistor destroys itself. This condition, of course, will not always occur if fixed-bias is used.

- The use of pnp transistors requires voltages of opposite polarity from those used for npn transistors. These voltages can be obtained by using a $-V_{CC}$. Another common method is to install the transistor with the emitter at $+V_{CC}$. No special negative voltage supply is required using this method. Remember, all BJT transistors, including pnp types, need to be biased the same.

- By now you can begin to see the value of properly understanding the conditions of saturation and cutoff. So many troubles can be analyzed by the three voltage measurements, V_C, V_B, and V_E. In your electronic lab, build these circuits. Simulate the troubles outlined in this chapter and see for yourself the voltages that will result. Your troubleshooting skills will increase greatly by this exercise.

CHAPTER 5 QUIZ

Student Name _____ Date _____

1. The purpose of biasing is to establish a proper Q-point.
   a. true
   b. false

2. A transistor is operating in a linear fashion at saturation.
   a. true
   b. false

3. The base bias circuit arrangement provides poor stability because its Q-point varies widely with β_{DC}.
   a. true
   b. false

4. Collector-feedback bias provides very poor stability with negative feedback from collector to base.
   a. true
   b. false

5. The formula $\beta_{DC} = I_B/I_C$ is correct.
   a. true
   b. false

6. What is the Q-point for a fixed-bias transistor with $I_B = 75\ \mu A$, $\beta_{DC} = 100$, $V_{CC} = 20$ V, and $R_C = 1.5$ kΩ?
   a. $V_C = 0$ V
   b. $V_C = 20$ V
   c. $V_C = 8.75$ V
   d. $V_C = 11.25$ V

7. Ideally, for linear operation, a transistor should be biased so that the Q-point is
   a. near saturation.
   b. near cutoff.
   c. where I_C is maximum.
   d. is half-way between cutoff and saturation.

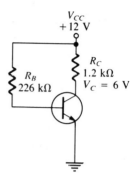

FIGURE 5-1

8. Refer to Figure 5-1. The value of I_B is
   a. 53 μA.
   b. 50 μA.
   c. 50 mA.
   d. 53 mA.

9. Refer to Figure 5-1. The value of I_C is
   a. 10 μA.
   b. 10 mA.
   c. 5 mA.
   d. 50 mA.

10. Refer to Figure 5-1. The value of β_{DC} is
    a. 5.3.
    b. 53.
    c. 94.
    d. 100.

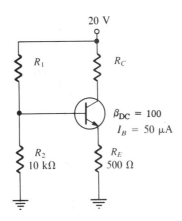

FIGURE 5-2

11. Refer to Figure 5-2. Determine I_C.
   a. 5 μA
   b. 5 mA
   c. 0 mA
   d. 10 mA

12. Refer to Figure 5-2. Assume that $I_C \cong I_E$. Find V_E.
   a. 5 V
   b. 10 V
   c. 15 V
   d. 2.5 V

13. Refer to Figure 5-2. Assume $I_C \cong I_E$. Determine the value of R_C that will allow V_{CE} to equal 10 V.
   a. 1 kΩ
   b. 1.5 kΩ
   c. 2 kΩ
   d. 2.5 kΩ

14. Refer to Figure 5-2. Calculate the current I_2.
   a. 32 mA
   b. 3.2 mA
   c. 168 μA
   d. 320 μA

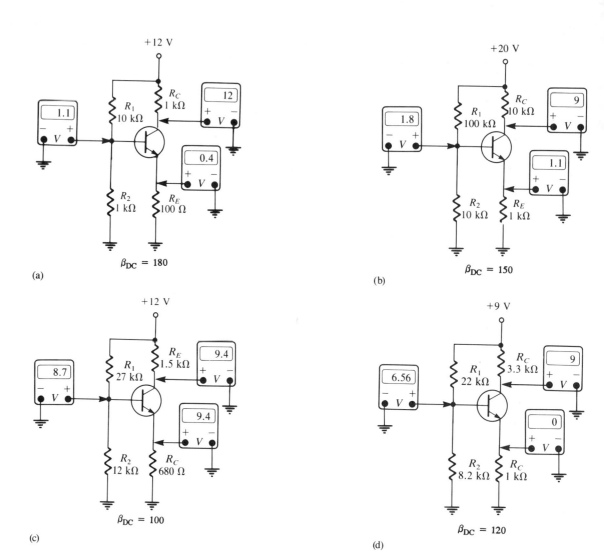

FIGURE 5-3

15. Refer to Figure 5-3(a). The most probable cause of trouble, if any, from these voltage measurements would be
    a. the base-emitter junction is open.
    b. R_E is open.
    c. a short from collector to emitter.
    d. no problems.

16. Refer to Figure 5-3(b). The most probable cause of trouble, if any, from these voltage measurements is
    a. the base-emitter junction is open.
    b. R_E is open.
    c. a short from collector to emitter.
    d. no problems.

17. Refer to Figure 5-3(c). The most probable cause of trouble, if any, from these voltage measurements is
    a. the base-emitter junction is open.
    b. R_E is open.
    c. a short from collector to emitter.
    d. no problems.

18. Refer to Figure 5-3(d). The most probable cause of trouble, if any, from these voltage measurements is
    a. the base-emitter junction is open.
    b. R_E is open.
    c. a short from collector to emitter.
    d. no problems.

19. The most stable biasing technique used is the
    a. voltage-divider bias.
    b. base-bias.
    c. emitter-bias.
    d. collector-bias.

20. At saturation the value of V_{CE} is nearly _____, and I_C = _____.
    a. zero, zero
    b. V_{CC}, $I_{C(sat)}$
    c. zero, $I_{C(sat)}$
    d. V_{CC}, zero

REVIEW OF KEY POINTS IN CHAPTER 6

SMALL-SIGNAL BIPOLAR AMPLIFIERS

SMALL-SIGNAL AMPLIFIER OPERATION

- A **small-signal amplifier** is used to amplify small voltages such as the ac input signal from a microphone, fm tuner, or tv antenna, which could have a very low p-p voltage. The purpose of small-signal amplifiers is to amplify the signal so that the amplitude is much larger.

- All amplifiers using BJT transistors need to be correctly biased. This means that the DC values should be correct.

- A **coupling capacitor** is used to couple the input signal to the amplifier input. The output of the amplifier also can use coupling capacitors to transfer the signal from the amplifier to the next circuit or stage.

- Coupling capacitors act as a very low reactance to the ac signal, thus passing the ac easily. At the same time, the DC voltage levels on each side of the coupling capacitor are isolated from each other. This preserves the DC bias levels.

- It is important to look at labels. Terms with capital subscripts are DC values: I_C, V_{CE}, and I_E. Notation in lowercase subscripts is used to indicate fixed ac values: I_c, V_{ce}, and I_e. Instantaneous ac values are all lower case: i_c, v_{ce}, and i_e.

TRANSISTOR AC EQUIVALENT CIRCUITS

- A **common-emitter** amplifier has the signal input on the base and the output at the collector.

- A **common-base** amplifier uses the emitter as the input and the collector as the output.

- The third type, **common-collector**, has the input on the base and the output at the emitter.

- The **h-parameter** is used by manufacturers to specify ac characteristics of the transistor.

- The **forward current gain** of a common-emitter amplifier is called $h_{fe}(\beta_{ac})$.

- Other common emitter h-parameters are **input resistance**, h_{ie}, and **voltage feedback ratio**, h_{re}.

- A common-emitter amplifier has a forward current gain of $h_{fe} = I_c/I_b$.

- Another useful set of parameters is the **r-parameters**. Commonly used parameters are alpha, $(\alpha_{ac}) = (I_c/I_e)$, beta, $(\beta_{ac}) = (I_c/I_b)$, and r'_e (ac emitter resistance).

- The ac emitter resistance can be estimated by this formula: $r'_e = 25 \text{ mV}/I_E$.

- The value of β_{DC} is not always equal to h_{fe} or β_{ac}.

COMMON-EMITTER AMPLIFIERS

- The DC analysis of the common-emitter amplifier was covered in Chapter 4.

- The ac analysis replaces all capacitors with a short, because a capacitor should have little reactance at ac frequencies. Remember, $X_C = \dfrac{1}{2\pi fC}$.

- The input impedance, which is low, seen by a source looking at the base is $R_{in(base)} = \beta_{ac} r'_e$.

- $r'_e \cong 25 \text{ mV}/I_E$, which is the ac emitter resistance.

- The total input impedance to the circuit is the parallel combination $R_{in(tot)} = R_1 \| R_2 \| R_{in(base)}$.

- The output resistance is high. $R_{out} \cong R_C$

- The output impedance is approximately equal to the collector resistor: $R_{out} \cong R_C$.

- Voltage gain of an unloaded common-emitter amplifier is found by $A_v = R_C/r'_e$.

- The addition of an emitter bypass capacitor will increase the voltage gain. The voltage gain formula is $A_v = R_c/r'_e$.

- Adding a load to an amplifier will decrease the voltage gain.

- A common-emitter amplifier has a **180°** phase shift between the input and output voltages.

- Current gain of a common-emitter amplifier is β_{ac}, or can be calculated using $A_i = I_c/I_s$.

- The power gain is always the product of voltage and current gains. $A_p = A_i A'_v$

- The overall voltage gain is $A'_v = \left(\dfrac{V_b}{V_s}\right) A_v$

COMMON-COLLECTOR AMPLIFIERS

- A common-collector is often called an **emitter-follower**.

- The voltage gain is approximately 1.

- Input impedance is high. The formula, $\mathbf{R_{in(base)}} = \beta_{ac}\mathbf{R_E}$, gives an approximate value for $R_{in(base)}$.

- The output impedance is very low. The formula to calculate the output impedance is $\mathbf{R_{out}} = \mathbf{R_s}/\beta_{ac} \parallel \mathbf{R_E}$.

- Since $A_v = 1$, the current gain $= \beta_{ac}$.

- Output voltage is in-phase or has a **0°** phase shift with the input voltage.

- A **Darlington Pair** is two direct-coupled common-collector amplifiers. The input impedance is very high.

- The current gain of a Darlington Pair is the product of the two values of β_{ac}. $\beta_{Darl} = \beta_{ac1} \beta_{ac2}$

- The input resistance of a Darlington Pair is $\mathbf{R_{in}} = \beta_{ac1} \beta_{ac2} \mathbf{R_E}$.

COMMON-BASE AMPLIFIERS

- The voltage gain is the same as for the common-emitter. $A_v = R_c/r'_e$

- Input impedance is low. $R_{in(emitter)} \cong r'_e$.

- The output impedance is about equal to R_C.

- Current gain is about 1.

- Common-base amplifiers have a power gain of about A_v. $A_p = A_v$

- The output voltage is in-phase or has a **0°** phase shift with the input voltage.

MULTISTAGE AMPLIFIERS

- Several amplifiers are often connected together to provide more voltage gain or current gain. This arrangement is called **cascaded stages**. Each amplifier is a separate stage.

- The overall voltage gain of a cascaded set of stages is found by $A'_v = A_{v1}A_{v2}A_{v3} \cdots A_{vn}$.

- Voltage gain is often expressed in decibels (dB). **(dB)A_v = 20 log A_v**

- Multistage amplifier voltage gain in dB is found by adding the voltage gains in dB.

- The second stage of an amplifier always loads the previous stage. This results in a lower first-stage voltage gain.

- Multistage amplifiers can be **direct-coupled**. This means that there is no coupling capacitor. The collector of one stage is directly connected to the base of the next.

- Direct-coupled amplifiers have a very good low-frequency response. They will amplify down to 0 Hz (DC).

TROUBLESHOOTING

- If you apply a known signal voltage to the input of a multistage amplifier, you can signal-trace through the amplifier using an oscilloscope. If you come to a portion of the circuit where the signal is missing or low in amplitude, the problem is in that part of the circuit.

- The DC bias voltage levels can also be checked with a DVM. Your understanding of these bias levels will indicate if an error is present.

TECHNICAL TIPS

- When considering the response of an amplifier to an ac signal, a capacitor is effectively a short. This makes the analysis much easier. Of course, if the capacitor is a short, it means that the capacitive reactance X_C is just a few ohms at the frequencies involved. At audio frequencies the capacitor might have to be quite large to give a low value of X_C.

- A common-emitter amplifier will often use an emitter bypass capacitor to increase the voltage gain. One way of looking at this is that since the capacitor acts as a short to ac, the emitter is at ac ground potential. The capacitor dismisses the effect of the emitter resistor R_E from the voltage gain formula. The emitter resistor is still used to provide the correct and stable bias voltages. The lower the frequency of operation, the larger the emitter bypass capacitor must be.

- There are many situations when it is desirable to approximate the input impedance of a common-emitter amplifier.

- A similar approximation for the input impedance of a common-base amplifier is also desired.

- The characteristics of all three BJT amplifier configurations are very important for you to know. For example, you need an amplifier with a very high input impedance to use as a buffer amplifier. With your knowledge of these characteristics, you would select a common-collector amplifier. You might also call it an emitter-follower.

- Practice following a signal through an amplifier in your electronic lab. Use the oscilloscope and see that the signal voltage on either side of a coupling capacitor is the same. Do the same by placing the oscilloscope lead on the bypassed emitter. The ac signal will indeed be very small. This is as it should be. You might even have to turn up the input gain of the scope to see any signal at all.

CHAPTER 6 QUIZ

Student Name _____ Date _____

1. A common-emitter amplifier has the advantages of good voltage, current, and power gain, but the disadvantage of a relatively low input impedance.
   a. true
   b. false

2. A Darlington Pair provides a very low input impedance.
   a. true
   b. false

3. A common-collector amplifier has a high input impedance, a good current gain, and a voltage gain of 1.
   a. true
   b. false

4. The common-base amplifier has a good voltage gain, a low input impedance, and a high current gain.
   a. true
   b. false

5. The total voltage gain of a multistage amplifier is the product of the individual stage gains.
   a. true
   b. false

6. The DC emitter current of a transistor is 8 mA. What is the value of r'_e?
   a. 320 Ω
   b. 13.3 kΩ
   c. 3.125 Ω
   d. 5.75 Ω

51

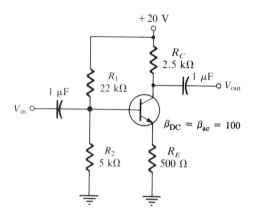

FIGURE 6-1

7. Refer to Figure 6-1. Calculate the value of V_B.
   a. 5 V
   b. 3.7 V
   c. 20 V
   d. 3 V

8. Refer to Figure 6-1. Find the value of I_E.
   a. 2 mA
   b. 4 mA
   c. 5 mA
   d. 6 mA

9. Refer to Figure 6-1. Determine the value of V_C.
   a. 20 V
   b. 10 V
   c. 5 V
   d. 0 V

10. Refer to Figure 6-1. Find the value of $R_{in(base)}$.
    a. 220 Ω
    b. 416.7 Ω
    c. 940 Ω
    d. 100.8 Ω

11. Refer to Figure 6-1. Calculate the value of $R_{in(tot)}$.
    a. 208.3 Ω
    b. 417 Ω
    c. 912 Ω
    d. 127 Ω

12. Refer to Figure 6-1. Determine the value of A_v.
    a. 166
    b. 720
    c. 871
    d. 600

13. Refer to Figure 6-1. If an emitter bypass capacitor was installed, determine the value of $R_{in(base)}$.
    a. 416.7 Ω
    b. 5 kΩ
    c. 4.07 kΩ
    d. 500 Ω

14. Refer to Figure 6-1. If an emitter bypass capacitor was installed, calculate the value of $R_{in(tot)}$.
    a. 50 Ω
    b. 175 Ω
    c. 208.3
    d. 500 Ω

15. Refer to Figure 6-1. If an emitter bypass capacitor was installed, what would the new A_v be?
    a. 4.96
    b. 125
    c. 398
    d. 600

16. An emitter-follower amplifier has an input impedance of 107 kΩ. The input signal is 12 mV. The approximate output voltage is
    a. 8.92 V.
    b. 112 mV.
    c. 12 mV.
    d. 8.9 mV.

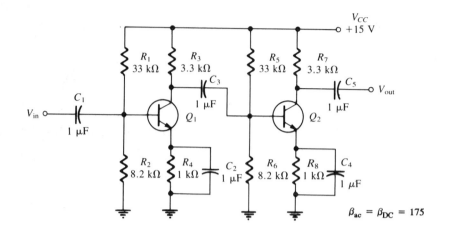

FIGURE 6-2

17. Refer to Figure 6-2. You note in servicing this amplifier that the output signal at V_{out} is reduced from normal. The problem could be caused by
    a. an open C_3.
    b. an open C_2.
    c. an open base-emitter of Q_2.
    d. a shorted C_2.

18. Refer to Figure 6-2. The output signal from the first stage of this amplifier is 0 V. The trouble could be caused by
    a. an open C_4.
    b. an open C_2.
    c. an open base-emitter of Q_1.
    d. a shorted C_4.

19. A Darlington Pair amplifier has
    a. high input impedance and high voltage gain.
    b. low input impedance and low voltage gain.
    c. a voltage gain of about 1 and a low input impedance.
    d. a low voltage gain and a high input impedance.

20. You have a need to apply an amplifier with a very high power gain. Which of the following would you choose?
    a. common-collector
    b. common-base
    c. common-emitter
    d. emitter-follower

REVIEW OF KEY POINTS IN CHAPTER 7

POWER AMPLIFIERS

CLASS A AMPLIFIERS

- A **class A amplifier** is biased so that the collector current I_C is flowing the entire cycle. The conducting angle is 360°.

- To obtain the maximum output signal amplitude, the Q-point should be biased in mid-range of the ac load line.

- If the Q-point is not centered, the signal peaks will enter the distortion range near cutoff or saturation.

- The load line, as plotted on a transistor's characteristic curves, runs between saturation current and cutoff.

- Saturation current is found by $I_{C(sat)} = V_{CC}/(R_C + R_E)$.

- At cutoff, $V_{CE} = V_{CC}$ and $I_C = 0$.

- Power gain A_p is found by $A_p = A_i A_v$.

- Output power from a power amplifier is found by $P_{out} = V_{ce}I_c$. These are ac values.

- The input DC power is found by $P_{DC} = V_{CC}I_C$.

- Efficiency of an amplifier is the ratio of input DC power to output ac power.
  Eff (%) = $(P_{out}/P_{DC})100$

- The maximum efficiency of a class A amplifier is 25%.

CLASS B AND AB PUSH-PULL AMPLIFIERS

- A **class B amplifier** is biased so that it will conduct during one-half the input cycle. This represents a 180° conduction angle.

- The class B amplifier is usually biased at 0 V. The positive swing of the signal will turn on the transistor.

- There is considerable distortion present since the output current only conducts 180° of the cycle.

- Push-pull amplifiers are connected so that one transistor is conducting during the positive alternation, and the other during the negative alternation.

- The distortion that occurs when the signal voltages cross zero and change polarity is called **crossover distortion**.

- The maximum efficiency of a class B amplifier is **78.5%**.

- To help eliminate crossover distortion, the **class B** push-pull amplifier is modified to **class AB**. This is done primarily with diodes in the bias circuit which will eliminate the crossover distortion.

- Class AB is biased between classes A and B near cutoff. It has a conduction angle of **200°**.

- The maximum efficiency of a class AB amplifier is **70%**.

CLASS C AMPLIFIERS

- A **class C amplifier** is biased below cutoff. The conduction angle may be considerably less than 180°; it is usually **90°**.

- The value of V_{BE} is biased negative so that a large signal swing will be required to turn on the transistor for just a short time during each cycle.

- This amplifier has **extreme** distortion. It is used with a tank circuit in tuned amplifiers. The tank operation will smooth the output voltage into a reasonable sine wave.

- The maximum theoretical efficiency is almost **100%**.

- A clamper is sometimes used to provide the negative bias required for this amplifier.

TECHNICAL TIPS

- You have been studying small-signal amplifiers. These amplifiers increase the amplitude of the signal voltage. Voltage alone will not supply appreciable power to drive a loudspeaker above a whisper. This application requires **power**. Power is $P = VI$. Now we need an amplifier that will increase the current levels. This, combined with our already adequate voltage levels, will provide us with the power needed to drive a larger load. Power amplifiers are usually current amplifiers. Some power amplifiers also provide voltage gain.

- When troubleshooting a power amplifier, an indication of trouble can be determined by measuring the DC voltage on the output of the transistor. If this voltage is not near the center of the load line, then possible distortion of the output signal could be the result. Another route is to look at the output of the amplifier with an oscilloscope. Clipping the output signal might be an indication of trouble in the DC bias voltages.

- The various classes of amplifier operation exemplify the compromises needed in electronic circuits. **Class A** offers the least distortion but the lowest efficiency. **Class B** has bad distortion but is considerably more efficient. **Class AB** has moderate distortion and efficiency. **Class C** is, of course, the most efficient, but the output signal needs a tank circuit to make it usable in most cases.

- Push-pull amplifiers require two transistors. Note that one is a **pnp** and the other is an **npn**. This combination is called a complementary amplifier. Complementary means that the transistors respond to opposite polarity voltages. There are other versions of the complementary amplifiers called complementary-symmetry amplifiers. These circuits use the npn and pnp transistors but have a + and − V_{CC} supply. The amplifier is then symmetrical about ground potential. These amplifiers are very common at audio frequencies.

- As you have learned, the DC bias voltages keep an amplifier operating correctly. Power amplifiers by their nature conduct large amounts of current. This means heat. Heat sinks are used to conduct the heat away from the transistor to protect it. These heat sinks act similarly to the radiator of your car. The more area exposed to air, the cooler the device will operate. In push-pull amplifiers, the biasing diodes are often mounted on the same heat sink as the output transistor. This allows the characteristics of both sets of pn junctions (in the transistor and in the diode) to change together at about the same rate. This keeps the DC bias voltages about correct, even with an increase in temperature.

CHAPTER 7 QUIZ

Student Name _____ Date _____

1. A class A amplifier conducts 180° of the cycle.
   a. true
   b. false

2. Class B amplifiers are usually operated in push-pull to obtain an output that is a near replica of the input signal.
   a. true
   b. false

3. The class C amplifier is biased below cutoff.
   a. true
   b. false

4. The least efficient amplifier among all classes is
   a. class B.
   b. class A.
   c. class AB.
   d. class C.

5. Class B amplifiers are usually zero-biased.
   a. true
   b. false

6. A class A amplifier has a voltage gain of 30 and a current gain of 25. What is the power gain?
   a. 30
   b. 25
   c. 1.2
   d. 750

7. You have an application for a power amplifier to operate at radio frequencies. The most likely choice would be a _____ amplifier.
   a. class A
   b. class B
   c. class C
   d. class AB

8. A class A amplifier with R_C = 3.3 kΩ and R_E = 1.2 kΩ has a V_{CC} = 20 V. Find $I_{C(sat)}$.
   a. 4.4 mA
   b. 6.1 mA
   c. 16.7 mA
   d. 20 mA

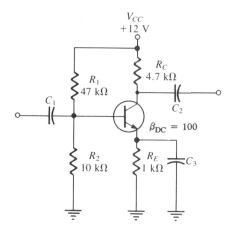

FIGURE 7-1

9. Refer to Figure 7-1. The DC voltage on the collector, V_C, is
   a. 5.4 V.
   b. 6.6 V.
   c. 12 V.
   d. 0 V.

10. Refer to Figure 7-1. This amplifier is operating as a _____ amplifier.
    a. class A
    b. class B
    c. class AB
    d. class C

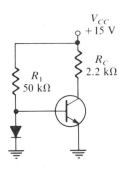

FIGURE 7-2

11. Refer to Figure 7-2. This amplifier is operating as a _____ amplifier.
    a. class A
    b. class B
    c. class AB
    d. class C

12. Refer to Figure 7-2. The approximate voltages on the base, collector, and emitter, respectively, are
    a. 0.7 V, 6.8 V, 0 V.
    b. 0 V, 0 V, 0 V.
    c. 0.7 V, 15 V, 0 V.
    d. 0.7 V, 0 V, 15 V.

13. Refer to Figure 7-2. The maximum efficiency of this amplifier is
    a. about 25%.
    b. about 78%.
    c. about 70%.
    d. about 100%.

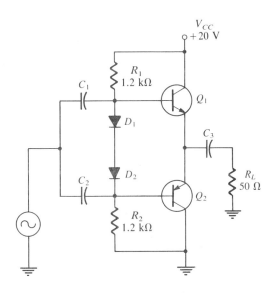

FIGURE 7-3

14. Refer to Figure 7-3. Determine V_{B1}.
    a. 0 V
    b. 0.7 V
    c. 9.3 V
    d. 10.7 V

15. Refer to Figure 7-3. Calculate V_{B2}.
    a. 0 V
    b. 0.7 V
    c. 9.3 V
    d. 10.7 V

16. Refer to Figure 7-3. You have an oscilloscope across R_L and it shows a zero signal voltage. The problem might be that
    a. C_3 is open.
    b. BE_1 is open.
    c. BE_2 is open.
    d. R_1 is open.

17. Refer to Figure 7-3. You find that this amplifier only shows the negative alternation at the output. The possible trouble might be that
    a. C_3 is shorted.
    b. BE_1 is open.
    c. BE_2 is open.
    d. R_1 is open.

18. Refer to Figure 7-3. You find that there is no output signal. You measure the DC voltage of Q_1 emitter and find it equal to 0 V. The trouble might be
    a. D_1 is shorted.
    b. D_2 is shorted.
    c. R_1 is open.
    d. no trouble, everything is normal.

19. Refer to Figure 7-3. You find that there is an input signal on the base of Q_1 and Q_2. However, there is no output signal. You then measure the DC voltages on Q_2 and find them to be all 0 V. The possible trouble might be
    a. C_3 is shorted.
    b. C_1 is open.
    c. R_L is shorted.
    d. V_{CC} is 0 V.

20. A class C amplifier has a tank circuit in the output. The amplifier is conducting only 28°. The output voltage is
    a. 0 V.
    b. a DC value equal to V_{CC}.
    c. a sine wave.
    d. a square wave with a frequency determined by the tank.

REVIEW OF KEY POINTS IN CHAPTER 8

FIELD-EFFECT TRANSISTORS AND BIASING

THE JUNCTION FIELD-EFFECT TRANSISTOR (JFET)

- The **junction field-effect transistor (JFET)** operates as a **voltage-controlled device**.

- The JFET is controlled by a voltage (V_{GS}). This voltage directly controls the **drain current (I_D)**.

- Recall that the BJT is controlled by the flow of base current; hence the BJT is a current-controlled device.

- There are two types of JFETs. One is an **n-channel** JFET and the other is a **p-channel** JFET.

- The JFET is a normally on device. It is biased so I_D flows through the channel.

- A JFET is always operated with the gate-source pn junction **reverse-biased**.

- The JFET's gate current I_G is practically zero.

- In an n-channel JFET, I_D decreases as V_{GS} becomes more negative because the channel becomes more narrow. When V_{GS} becomes more positive, I_D increases. Remember that the value of V_{GS} must remain negative, thus reverse-biasing the gate-source junction.

JFET CHARACTERISTICS AND PARAMETERS

- The **pinch-off voltage (V_P)** is the value of V_{DS} where the channel becomes saturated and an increase in I_D will not occur.

- The value of I_D at pinch-off is called I_{DSS}.

- Most manufacturers' data sheets do not specify a value for V_P. Rather, they specify $V_{GS(off)}$. These two values are equivalent to each other: $V_P = V_{GS(off)}$.

- The relationship between I_D and I_{DSS} can be determined by the formula:
  $I_D = I_{DSS}(1 - V_{GS}/V_{GS(off)})^2$.

- The JFET **forward transconductance, g_m**, is the change in drain current for a given change in V_{GS}.

- The **input resistance, R_{IN}**, of a JFET is found by V_{GS}/I_{GSS}. This averages many thousands of MΩ.

JFET BIASING

- A JFET must be biased so that the pn junction between the gate and source is reverse-biased. This is easily accomplished by making $V_G = 0$ V. The drain current will then produce a drop across the source resistor, R_S. V_{GS} will now be negative, thus biasing the pn junction correctly.

- For an n-channel JFET, the value of V_{GS} can be found by $V_{GS} = -I_D R_S$ when $V_G = 0$ V.

- To achieve linear operation, it is common to set the operating point midway between the maximum and minimum drain current. $I_D = I_{DSS}/2$. This is called **midpoint biasing**.

- A midpoint bias approximation for V_{GS} is $V_{GS} = V_{GS(off)}/4$. These two values for I_D and V_{GS} will usually produce linear operation.

- The values of R_S and R_D can be found by Ohm's Law. $R_S = V_{GS}/I_D$ and $R_D = (V_{DD} - V_D)/I_D$.

THE METAL OXIDE SEMICONDUCTOR FET (MOSFET)

- The **MOSFET (metal oxide semiconductor field-effect transistor)** has no pn junction. The gate is made of metal and is insulated from the source by a layer of S_iO_2.

- The MOSFET has a higher input resistance than the JFET.

- A MOSFET is available in two modes, **depletion (D)** and **enhancement (E)**.

- The **D-MOSFET** can operate with either a positive or negative value of V_{GS}. A negative value of V_{GS} operates the MOSFET in the depletion mode. A positive value operates the MOSFET in an enhancement mode.

- The depletion mode acts as a capacitor with one plate as the gate and the other plate as the channel. As the gate becomes more negative, the channel is depleted of electrons. This causes I_D to decrease.

- The enhancement mode also acts as a capacitor, but as V_{GS} becomes positive, electrons are attracted into the channel. This enhances the channel. As V_{GS} increases, I_D increases.

- The **E-MOSFET** operates only in the enhancement mode.

- The construction of the E-MOSFET is similar to the D-MOSFET, except that there is no channel. A positive value of V_{GS} enhances the channel and allows I_D to flow.

- The **V-MOSFET**, or **VMOS**, is an enhancement mode device that allows greater current to flow. The construction of the enhanced channel is wider and in the shape of a V.

- A V-MOSFET has a much higher current rating than a regular MOSFET. Frequency response is greatly improved also.

MOSFET BIASING

- A zero voltage on the gate of a D-MOSFET, with the source grounded, is a simple method of biasing.

- The value of V_{DS} for a D-MOSFET can be found by $V_{DS} = V_{DD} - I_{DSS}R_D$.

- E-MOSFETs require a positive V_{GS}. A voltage-divider bias arrangement can be used.

TECHNICAL TIPS

- The voltage gain A_V of a JFET is not as high as a BJT. A formula for calculating the voltage gain for a JFET amplifier is $A_v = g_m R_D$. The transconductance, g_m, is given in the data sheets.

- Biasing a JFET circuit is easy. Usually JFETs are self-biased. Since the gate current is zero, the addition of a source resistor will reverse bias the GS junction by producing a positive voltage at the source. This will make V_{GS} negative. The gate resistor, R_G, is selected to be a large value. This prevents the circuit from loading the source of the signal input. The input impedance of the JFET is very high.

- When it is necessary to set the bias values on a JFET circuit, a common method is to set $I_D = I_{DSS}/2$. This gives the drain current room to increase and decrease in response to signal changes. In addition, setting $V_{GS} = V_{GS(off)}/4$ will also place V_{GS} about the midpoint of its range between V_P and 0 V.

- VMOS (vertical metal oxide semiconductor transistor) is now competing successfully against the power BJT. MOSFETs have currents in the mA range, but VMOS can have currents in the ampere range. VMOS is also used as a high current switch in some applications.

- Always use special care in handling a MOSFET of any type. The metal-oxide gate insulator is very thin and has a very high resistance. The leakage current is measured as a few pA—not much current at all. This insulator is **very** susceptible to static charge. Make sure that the device remains in its shipping ring until it is soldered into the circuit board. Remember to solder with an iron with a good, solid connection to ground. This helps to bleed off any accumulated static charge.

- D-MOSFETs are used in amplifiers that require very high input resistance. Low noise is another good reason to use a MOSFET at radio frequencies. Some MOSFETs have dual gates so they can be used in circuits such as automatic gain control (AGC).

- An E-MOSFET is used mostly in switching circuits. An entire digital family of logic chips is based upon MOSFETs, called the CMOS logic family.

CHAPTER 8 QUIZ

Student Name _____ Date _____

1. The three FET terminals are called source, base, and drain.
    a. true
    b. false

2. The JFET operates with a reverse-biased pn junction (gate-to-source).
    a. true
    b. false

3. A VMOS device can handle higher power and voltage than a conventional MOSFET.
    a. true
    b. false

4. The E-MOSFET has no physical channel.
    a. true
    b. false

5. It is not necessary to exercise any particular care in handling a MOSFET.
    a. true
    b. false

6. An n-channel JFET has a $V_D = 6$ V. $V_{GS} = -3$ V. Find the value of V_{DS}.
    a. -3 V
    b. -6 V
    c. 3 V
    d. 6 V

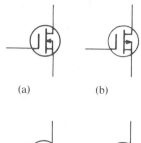

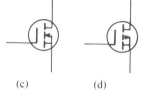

FIGURE 8-1

7. Refer to Figure 8-1. Identify the p-channel E-MOSFET.
    a. a
    b. b
    c. c
    d. d

8. Refer to Figure 8-1. Identify the n-channel D-MOSFET.
    a. a
    b. b
    c. c
    d. d

9. Refer to Figure 8-1. Identify the n-channel E-MOSFET.
    a. a
    b. b
    c. c
    d. d

10. Refer to Figure 8-1. Identify the p-channel D-MOSFET.
    a. a
    b. b
    c. c
    d. d

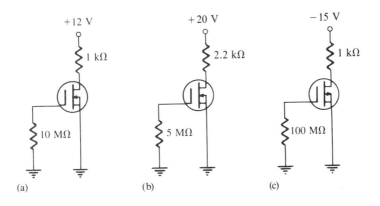

FIGURE 8–2

11. Refer to Figure 8-2(a). $I_D = 6$ mA. Calculate the value of V_{DS}.
    a. -6 V
    b. 6 V
    c. 12 V
    d. -3 V

12. Refer to Figure 8-2(b). $I_D = 6$ mA. Calculate the value of V_{DS}.
    a. 13.2 V
    b. 10 V
    c. 6.8 V
    d. 0 V

13. Refer to Figure 8-2(c). $I_D = 6$ mA. Calculate the value of V_{DS}.
    a. -9 V
    b. 9 V
    c. 6 V
    d. -3 V

14. A JFET data sheet specifies $V_{GS(off)} = -6$ V and $I_{DSS} = 8$ mA. Find the value of I_D when $V_{GS} = -3$ V.
    a. 2 mA
    b. 4 mA
    c. 8 mA
    d. none of these

15. A JFET data sheet specifies $V_{GS(off)} = -10$ V and $I_{DS} = 8$ mA. Find the value of I_D when $V_{GS} = -3$ V.
    a. 2 mA
    b. 1.4 mA
    c. 4.8 mA
    d. 3.92 mA

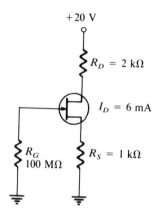

FIGURE 8–3

16. Refer to Figure 8–3. Determine the value of V_S.
   a. 20 V
   b. 8 V
   c. 6 V
   d. 2 V

17. Refer to Figure 8–3. Calculate the value of V_D.
   a. 20 V
   b. 8 V
   c. 6 V
   d. 2 V

18. Refer to Figure 8–3. Find the value of I_G.
   a. 6 mA
   b. 4 mA
   c. 2 mA
   d. 0 mA

19. Refer to Figure 8–3. Determine the value of V_{GS}.
   a. −20 V
   b. −8 V
   c. −6 V
   d. −2 V

20. Refer to Figure 8–3. Calculate the value of V_{DS}.
   a. 0 V
   b. 2 V
   c. 4 V
   d. −2 V

REVIEW OF KEY POINTS IN CHAPTER 9

SMALL-SIGNAL FET AMPLIFIERS

SMALL-SIGNAL FET AMPLIFIER OPERATION

- Self-biased JFET amplifier circuits usually have a very large resistor, R_G, connected from the gate to ground. This keeps the DC voltage on the gate at 0 V.

- A signal, capacitively coupled to the gate, causes V_{GS} to become more or less negative.

- I_D increases in step or in phase with the input signal at the gate.

- The voltage at the drain, V_D, is out of phase with the input signal.

- A self-biased D-MOSFET has a similar circuit to the self-biased JFET. The insulated gate allows V_{GS} to be either positive or negative due to signal voltages.

- The E-MOSFET can be biased with a voltage-divider circuit. Remember, V_{GS} must be kept positive for drain current to flow in an n-channel MOSFET.

FET AMPLIFICATION

- In a JFET circuit ac analysis, capacitors will be replaced by shorts.

- Voltage gain is found by the formula $A_v = g_m R_d$.

- Adding an external source resistor will lower the voltage gain.

COMMON-SOURCE AMPLIFIERS

- The input to a common-source amplifier is at the gate. The output is at the drain.

- Biasing the circuit at the midpoint of the load line will provide a drain current found by $I_D = I_{DSS}/2$.

- The output voltage is calculated by $V_{out} = A_v V_{in}$.

- There is a phase inversion between V_{in} and V_{out}. This phase inversion is 180°.

- The input resistance to this circuit can be calculated by the formula
$R_{in} = (V_{GS}/I_{GSS}) \parallel R_G$.

- The common-source amplifier using D- or E-MOSFETs is similar to JFET circuits. Recall, however, that D-MOSFETs are usually biased at 0 V, while E-MOSFETs have a positive value of V_{GS}.

COMMON-DRAIN AMPLIFIERS

- This amplifier has the input at the gate and the output at the source.

- A common name for this circuit is **source-follower**.

- The voltage gain is found by $A_v = g_m R_S/(1 + g_m R_S)$.

- No phase inversion is found in this amplifier.

- The input resistance to this circuit can be calculated by the formula $R_{in} = (V_{GS}/I_{GSS}) \parallel R_G$.

TECHNICAL TIPS

- A positive-going signal applied to the gate of a JFET amplifier causes V_{GS} to become more positive. When this occurs, the gate is, in effect, opened; and more electrons, hence more current, flows. So as V_{GS} opens and closes the gate, more or less I_D flows. Remember that V_{GS} only becomes more or less negative. The bias must be correct.

- Applying a load to either the input or output of any circuit will always cause the voltage to decrease. This is shown repeatedly in the voltage gain formulas. At audio frequencies, the loading effect is not always critical. Radio frequencies present a different set of circumstances. The signal amplitude at high frequencies is usually very small. Because of this, it is usual to design circuits to have as little loading effect as possible.

- Using the value of R_G approximates the input impedance for a JFET amplifier. This will not give you an exact R_{in}, but in some cases this value will prove adequate.

- The voltage gain formula for a common-drain amplifier was given earlier. Calculations with this formula will give a voltage gain so close to 1 that for all practical purposes you can consider the voltage gain to be 1. You know the amplifier is called a source-follower. This means that the source follows the input signal, both in phase and amplitude.

- Troubleshooting FET amplifiers is similar to troubleshooting BJT amplifiers. Recall that coupling capacitors, if working properly, will offer no attenuation to the signal voltages. If the coupling capacitor is open or leaky, you will find a drop in signal across the capacitor. The same thinking will apply to an open or leaky bypass capacitor. The signal voltage at the source should be very close to zero. If not, then suspect the capacitor.

CHAPTER 9 QUIZ

Student Name _____ Date _____

1. The voltage gain of a common-source amplifier is found by the product of g_m and R_D.
   a. true
   b. false

2. There is no phase inversion between the gate and the drain voltages.
   a. true
   b. false

3. There is a 180° phase inversion between the gate and source voltages.
   a. true
   b. false

4. A load resistance connected to the output of an amplifier reduces the voltage gain.
   a. true
   b. false

5. Bypassing a source resistor reduces the voltage gain.
   a. true
   b. false

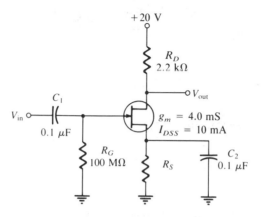

FIGURE 9-1

6. Refer to Figure 9-1. For midpoint biasing, I_D would be
   a. 10 mA.
   b. 7.5 mA.
   c. 5 mA.
   d. 2.5 mA.

7. Refer to Figure 9-1. Find the value of V_D.
   a. 20 V
   b. 11 V
   c. 10 V
   d. 9 V

8. Refer to Figure 9-1. If $V_{GS} = -6$ V, calculate the value of R_S that will provide this value.
   a. 2.2 kΩ
   b. 1.2 kΩ
   c. 600 Ω
   d. 100 Ω

9. Refer to Figure 9-1. The voltage gain is
   a. 1.2.
   b. 2.4.
   c. 4.4.
   d. 8.8.

10. Refer to Figure 9-1. If $V_{in} = 20$ mV p-p, what is the output voltage?
    a. 176 mV p-p
    b. 88 mV p-p
    c. 48 mV p-p
    d. 24 mV p-p

11. Refer to Figure 9-1. If $V_{in} = 1$ V p-p, the output voltage V_{out} would be
    a. undistorted.
    b. clipped on the negative peaks.
    c. clipped on the positive peaks.
    d. 0 V p-p.

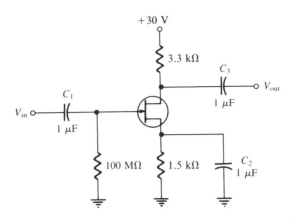

FIGURE 9-2

12. Refer to Figure 9-2. If $I_D = 4$ mA, $I_{DSS} = 16$ mA, and $V_{GS(off)} = -8$ V, find V_{DS}.
    a. 19.2 V
    b. −6 V
    c. 10.8 V
    d. 30 V

13. Refer to Figure 9-2. If $I_D = 4$ mA, find the value of V_{GS}.
    a. 10.8 V
    b. 6 V
    c. −0.7 V
    d. −6 V

14. Refer to Figure 9-2. If $g_m = 4000$ μS and a signal of 75 mV rms is applied to the gate, calculate the p-p output voltage.
    a. 990 mV
    b. 1.13 V p-p
    c. 2.8 V p-p
    d. 990 V p-p

15. Refer to Figure 9-2. The approximate value of R_{in} is
    a. 100 MΩ.
    b. 1.5 kΩ.
    c. 3.3 kΩ.
    d. 48 MΩ.

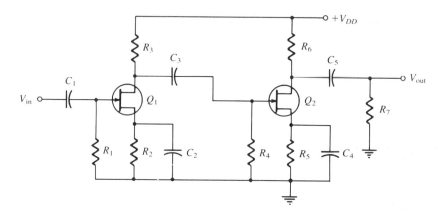

FIGURE 9-3

16. Refer to Figure 9-3. If C_4 opened, the signal voltage at the drain of Q_1 would
    a. increase.
    b. decrease.
    c. remain the same.
    d. distort.

17. Refer to Figure 9-3. If R_7 were to decrease in value, V_{out} would
    a. increase.
    b. decrease.
    c. remain the same.
    d. distort.

18. Refer to Figure 9-3. If C_2 shorted, V_{out} would
    a. increase.
    b. decrease.
    c. remain the same.
    d. distort.

19. Refer to Figure 9-3. If V_{in} was increased in amplitude a little, what would the signal voltage at the source of Q_2 do?
    a. increase
    b. decrease
    c. remain the same
    d. distort

20. Refer to Figure 9-3. If R_6 opened, the signal at the drain of Q_1 would
    a. increase.
    b. decrease.
    c. remain the same.
    d. distort.

REVIEW OF KEY POINTS IN CHAPTER 10

AMPLIFIER FREQUENCY RESPONSE

GENERAL CONCEPTS

- Coupling capacitors have a great effect on the low-frequency response of an amplifier.

- As the frequency of operation decreases, the X_C of the coupling capacitor increases. This causes an ac drop in signal amplitude at low frequencies.

- Bypass capacitors also affect the low-frequency response. If the X_C of the bypass capacitor is high, then the emitter (for example) will no longer be at ac ground.

- Internal transistor capacitances affect the high frequencies of operation.

- As the frequency of operation increases, the internal capacitance between the collector and the base (C_{cb}) allows a negative feedback path with lower $X_{C(cb)}$ between the two elements.

- The internal capacitance of a JFET, C_{gd}, has the same effect on this type of transistor as C_{cb} in a BJT.

- **Miller's Theorem** states that the internal capacitance has a greater effect on the high-frequency response than the actual capacitance value would indicate: $C_{in(Miller)} = C_{bc}(A_v + 1)$.

THE DECIBELS

- Power gain can be expressed in decibels by $(dB)A_p = 10 \log A_p$.

- The voltage gain is calculated by $(dB)A_v = 20 \log A_v$.

- A positive voltage gain expressed in dB indicates a signal increase. A negative voltage gain expressed as $-dB$ indicates attenuation. These positive or negative dB values are in reference to 0 dB.

- The critical frequency of an amplifier (both upper and lower) is that frequency at which the voltage gain will be down 3 dB, or -3 dB.

- Power gain measurements usually use a unit called a **dBm (dB milliwatt)**.

- A dBm references the power gain measurement to a level of 1 mW.

GAIN ROLL-OFF

- A **ten-fold change** in frequency is called a **decade**.

- A two times increase or decrease in the value of a quantity such as frequency is called an **octave**.

- The dB voltage attenuation for each decade is -20 dB.

- The gain roll-off occurs at both the upper and lower frequencies of amplifier operation.

- Low-frequency roll-off is affected by coupling and bypass capacitors.

- High-frequency roll-off is largely affected by the internal capacitances of the BJT or JFET.

- **Phase shift** also occurs in an RC network such as is found in an RC-coupled amplifier.

- The phase shift in the output RC network is found by $\theta = \arctan(X_{C2}/(R_C + R_L))$, where C_2 is the output coupling capacitor.

- A direct-coupled amplifier has no coupling or bypass capacitors, and the gain stays uniform down to 0 Hz (DC).

- The high frequency roll-off is controlled by Miller's capacitance.

- A common name for the critical frequency, f_c, is **cutoff frequency**.

- The cutoff frequency occurs at the higher and lower frequencies, which are called **f_{ch}** and **f_{cl}**.

- **Bandwidth** is the range of frequencies between the higher and lower cutoff frequencies. $BW = f_{ch} - f_{cl}$

- As the operating frequency increases or decreases beyond cutoff, the gain decreases until $A_v = 0$ dB. This frequency is called **unity-gain frequency, f_T**.

- The f_T is called **gain-bandwidth product**. $f_T = A_{v(mid)}BW$

- The capacitances that affect the frequency response for JFET amplifiers are similar to those used in BJT amplifiers.

TECHNICAL TIPS

- The low frequency response of a JFET or BJT amplifier using coupling or bypass capacitors can be thought of as a low-pass filter. At f_c the gain is down 3 dB. This is the same response that you have in a low-pass filter. If you know the approximate value of R_{in} and the coupling capacitor C, you can find the value of f_c by $f_c = 1/2\pi R_{in}C$.

- High-frequency response is limited largely by the internal capacitance of the transistor. This capacitance, in conjunction with circuit resistances, acts as a high-pass filter. At very high frequencies, X_C becomes lower and bypasses the low frequencies to ground. This is high-pass action.

- The decibels, dB, is a convenient unit to use to express either very large or very small numbers. In electronics we deal with very large quantities, such as A_v. Small quantities might be the amount of signal lost in a TV transmission from Mars. A decibel is nothing more than the log of a ratio of two powers, two voltages, or two currents. Remember the basic formula for dBs: $dB = 10 \log A_p$. For ratios of voltage or current, change the 10 to a 20, like this: $dB = 20 \log A_i$ or $dB = 20 \log A_v$.

- An increase of power by a ratio of two represents a dB gain of 3 dB. If you cut the power in half, it is a -3 dB loss. Filters and amplifiers roll-off beyond the f_c at a usual rate of 20 dB per decade. Sometimes roll-off is measured in dB per octave. An octave is a doubling or halving of a frequency. A roll-off of 20 dB per decade is equal to a roll-off of 6 dB per octave.

- At the cutoff frequency, the output voltage is down -3 dB from the midpoint value. A very convenient number to remember is that -3 dB represents a decimal multiplier of 0.707. For example, if the output voltage from an amplifier is 10 V p-p at 1 kHz, then the output will be 0.707 × 10 V or 7.07 V p-p at the f_c.

CHAPTER 10 QUIZ

Student Name _____ Date _____

1. The coupling and bypass capacitors affect the high-frequency response of an amplifier.
   a. true
   b. false

2. The internal transistor capacitances affect the frequency response of amplifiers.
   a. true
   b. false

3. At the cutoff frequency the output is down 20 dB.
   a. true
   b. false

4. A decade of frequency change is a change by a factor of 10.
   a. true
   b. false

5. Bandwidth is the difference between f_{cl} and f_{ch}.
   a. true
   b. false

6. An amplifier has an output voltage of 7.6 V p-p at the midpoint of the frequency range. What is the output at f_c?
   a. 3.8 V p-p
   b. 3.8 V_{rms}
   c. 5.4 V_{rms}
   d. 5.4 V p-p

7. An amplifier has an input signal voltage of 0.054 mV. The output voltage is 12.5 V. The voltage gain in dB is
   a. 53.6 dB.
   b. 107.3 dB.
   c. 231 dB.
   d. 116 dB.

8. A certain amplifier has a bandwidth of 22.5 kHz with a lower cutoff frequency of 600 Hz. What is the value of f_{ch}?
   a. 600 Hz
   b. 22.5 kHz
   c. 23.1 kHz
   d. 21.9 kHz

9. An amplifier has an R_{in} = 1.2 kΩ. The coupling capacitor is 1 μF. Determine the approximate lower cutoff frequency.
   a. 133 Hz
   b. 1.33 kHz
   c. 13.3 kHz
   d. 133 kHz

10. An RC network has values of R = 1.2 kΩ and C = 0.22 μF. Find f_c.
    a. 3.79 kHz
    b. 1.89 kHz
    c. 603 Hz
    d. 60 Hz

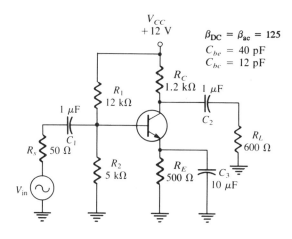

FIGURE 10–1

11. Refer to Figure 10–1. The capacitor C_1 affects
    a. high-frequency response.
    b. low-frequency response.
    c. midrange response.
    d. nothing.

12. Refer to Figure 10–1. The capacitor C_3 affects
    a. high-frequency response.
    b. low-frequency response.
    c. midrange response.
    d. nothing.

13. Refer to Figure 10–1. The capacitor C_{be} affects
    a. high-frequency response.
    b. low-frequency response.
    c. midrange response.
    d. nothing.

14. Refer to Figure 10-1. If R_L decreases in value, the output voltage will
    a. increase.
    b. decrease.
    c. remain the same.

15. Refer to Figure 10-1. The upper cutoff frequency of this amplifier is 22 kHz. The output at that frequency is 6.71 V p-p. What is the output voltage at 220 kHz?
    a. 9.49 V p-p
    b. 6.71 V p-p
    c. 0.671 V p-p
    d. 0.0671 V p-p

16. Refer to Figure 10-1. You measure an output voltage at the lower cutoff frequency of 3.42 V p-p. The output voltage at the upper cutoff frequency will be
    a. 2.42 V p-p.
    b. 3.42 V p-p.
    c. 6.84 V p-p.
    d. 6.84 V_{rms}.

17. Refer to Figure 10-1. The output voltage at f_{cl} = 12 mV. What is the output voltage at the midpoint frequency?
    a. 12 mV
    b. 12 mV p-p
    c. 16.97 mV
    d. 8.48 mV

18. Refer to Figure 10-1. You are attempting to determine the lower cutoff frequency of this amplifier in the lab. As you change the input frequency and measure the output signal, you must remember to
    a. set the oscilloscope to DC.
    b. maintain the input voltage constant.
    c. keep a constant temperature.
    d. watch for a change of β.

19. An RC network has a roll-off of 20 dB per decade. What is the total attenuation between the output voltage in the midrange of the pass-band as compared to the output voltage at a frequency of 10 times f_c?
    a. −3 dB
    b. −20 dB
    c. −23 dB
    d. −43 dB

20. A roll-off of 20 dB per decade is equivalent to a roll-off of _____ per octave.
    a. 3 dB
    b. 13 dB
    c. 12 dB
    d. 6 dB

REVIEW OF KEY POINTS IN CHAPTER 11

THYRISTORS AND OTHER DEVICES

THE SHOCKLEY DIODE

- A **thyristor** is the name of a family of four-layer semiconductor devices.

- Thyristors are devices such as the **Schockley diode, silicon-controlled rectifier (SCR), silicon-controlled switch (SCS), diac,** and **triac**.

- The Schockley diode has two terminals, the anode and cathode.

- Applying a positive voltage with respect to the cathode on the anode will cause the diode to conduct when this voltage exceeds the **forward breakover voltage ($V_{BR(F)}$)**.

- The Schockley diode will continue to conduct until the minimum **holding current (I_H)** value is reached. The diode will then cease to conduct.

- Schockley diodes are often used in **relaxation oscillators**.

SILICON-CONTROLLED RECTIFIER (SCR)

- The **silicon-controlled rectifier (SCR)** is another four-layer device with three terminals. These are the anode, cathode, and gate.

- The SCR is placed in the circuit in the forward-bias position. The anode is positive with respect to the cathode.

- If the forward breakover voltage is high enough, the SCR will conduct.

- Normally, to turn an SCR on, a positive pulse is applied to the gate. The gate current, I_G, will allow the SCR to fire.

- Once an SCR conducts, the only way to turn it off is to decrease the anode current below a value called the **holding current (I_H)**.

- Interrupting the anode current will stop the conduction. A switch in parallel with the SCR that can be actuated to reduce the current is another method of turning the SCR off.

SCR APPLICATIONS

- An SCR has many applications in the control of power.

- A common usage for an SCR is in half-wave power control of electric heaters, drill motors, or lamp dimmers, to name a few.

- A potentiometer in the gate circuit can adjust the turn-on time to any time within the first 90° of the positive ac alternation.

- As the conduction time or angle is increased, the average power delivered to the load will increase. This controls the temperature of a heater, for instance.

- When used on an ac application, the SCR is turned off automatically when the supply voltage nears zero. No conduction occurs during the negative alternation.

THE SILICON-CONTROLLED SWITCH (SCS)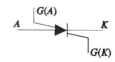

- The **silicon-controlled switch** is similar to the SCR, except there are two gates. These are the **anode gate** and the **cathode gate**.

- An SCS can be turned on by a positive pulse at the cathode gate, or by a negative pulse at the anode gate.

- Turning off an SCS can be accomplished by a positive pulse at the anode gate or a negative pulse at the cathode gate.

- An SCS has the advantage over the SCR of faster turn-off and the ability to turn the device off with a pulse. SCS current ratings are limited, however.

THE DIAC AND TRIAC

- The **diac** is a four-layer device that will conduct in both directions.

- Conduction in the diac will occur when the breakover voltage is reached. Either polarity will cause conduction.

- The diac turns off when the current decreases below the holding value.

- Another four-layer device is the **triac**. This is essentially two SCRs back-to-back. There are one gate terminal and two anode terminals.

- A positive pulse will trigger the triac into conduction on either the positive or negative alternation.

- The triac, similar to the other four-layer devices, is turned off by reducing the current below the holding value.

- The triac is more efficient than the SCR on ac, because the triac conducts on both alternations. This will deliver two times the power to the load.

- Conduction time can be easily controlled in a manner similar to the SCR.

THE UNIJUNCTION TRANSISTOR (UJT)

- A three-terminal device used to control current is the **unijunction transistor (UJT)**. The three terminals are the emitter, base 1, and base 2.

- In the symbol for the UJT, the emitter arrow points toward the base 1 terminal.

- A positive pulse on the emitter will cause the UJT to conduct. The device will continue to conduct until the emitter voltage is reduced to a value known as the **valley point**. The UJT will then shut off.

- A UJT is used to trigger other devices, such as the SCR. A relaxation oscillator is sometimes made using a UJT.

THE PROGRAMMABLE UNIJUNCTION TRANSISTOR (PUT)

- A **PUT** is similar in construction to an SCR, except the gate is on the anode side of the PUT.

- When the anode voltage exceeds the gate voltage by about 0.7 V, the PUT conducts. Lowering this voltage will turn off the PUT.

- The gate voltage may be set easily with a voltage divider, thus making the turn-on voltage programmable.

THE PHOTOTRANSISTOR

- A **phototransistor** is a transistor with a light-activated collector-base junction.

- A window allows light to enter the transistor. The greater the incident light entering, the higher the value of collector current.

- The phototransistor behaves in a similar manner to a BJT, except there is no base connection and the method of biasing is different (light rather than voltage).

- A **photodarlington** is a darlington arrangement of transistors with the driver transistor a phototransistor. Large current gains can be achieved.

- These phototransistors are widely used in control applications requiring light input. Some examples are relay control and burglar alarms.

THE LIGHT-ACTIVATED SCR (LASCR)

- The **LASCR** is an SCR that can be triggered by incident light through a window in the device.

- Two triggering modes can be used with the **LASCR**: normal gate current and light.

- The output **LASCR** load current circuit is electrically isolated from the input light source.

OPTICAL COUPLERS

- Many electronic applications need to isolate one circuit with its voltages or grounds from another. A device to accomplish this is the **optical coupler**.

- A typical optical coupler consists of an **LED** and one type of photosensitive device sealed in a light-proof package.

- The light emitted by the **LED** is received by the light-sensitive device. Complete electrical isolation is achieved. The only path between the input and output is an optical signal.

- The receiving device can be a phototransistor, a photodarlington, a photodiode, or a **LASCR**.

- A popular application of these optical couplers is the connection of electronic equipment to telephone systems.

- Optical couplers are also used to transfer an electrical signal to a **fiber optic cable** for transmission to other locations.

TECHNICAL TIPS

- The SCR is useful in the speed control of small ac motors, and more important, in DC motor control systems. They are quite efficient in handling large amounts of power as well. For example, many railway traction systems use very large SCRs to provide DC to the traction motors. Some diesel electric locomotives use SCRs with ratings of 4000 V at 4000 amperes. Each locomotive has a bank of many of this type of SCR. The SCR for this application is quite small, about 4" in diameter and 2" thick. They are very expensive.

- A useful device to act as a trigger for an SCR or triac is the diac. The diac has the ability to conduct in both directions, and the large surge of current when a diac fires is often used to trigger another four-layer device.

- Control of ac currents is the real field for the triac. The ability of the triac to conduct on both alternations gives this device an advantage in efficiency over the SCR in small-motor control. The power contained in the negative alternation is not wasted as it would be in an SCR application.

- In order to offer a minimum average voltage to a light or motor, it is necessary to control either the SCR or triac over more than 90° of each alternation. This is commonly done by a simple RC network. The time needed to charge a capacitor through a resistor can delay the application of the trigger voltage to the gate of the device by up to 180°. This gives a most flexible circuit arrangement.

- Testing four-layer devices is somewhat more difficult than testing transistors or diodes. The higher voltages required for breakover make the ordinary ohmmeter unsatisfactory. A diac may be tested by placing it in series with a resistor and observing the voltage across the diac as the applied voltage is increased. If the diac fires at the correct voltage, then it is probably all right. An SCR is often tested in a simple circuit to measure the gate current when the SCR fires. A momentary trigger pulse makes troubleshooting these devices more difficult. It is common to replace the device and see if the trouble is cleared up.

CHAPTER 11 QUIZ

Student Name _____ Date _____

1. Thyristors are four-layer semiconductor devices.
   a. true
   b. false

2. The SCR is a device that can be triggered off by a pulse applied to the gate.
   a. true
   b. false

3. The diac can conduct current in either direction and is turned on when a breakover voltage is exceeded.
   a. true
   b. false

4. The PUT can be programmed to turn on at a desired voltage level.
   a. true
   b. false

5. The triac can be turned on or off by a pulse at the gate.
   a. true
   b. false

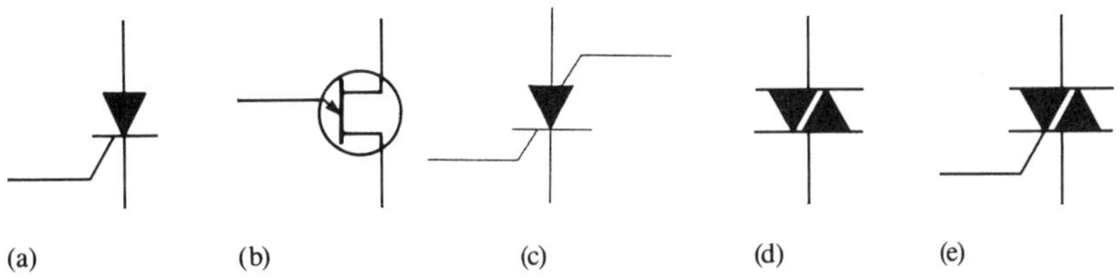

FIGURE 11-1

6. Refer to Figure 11-1. Identify the triac.
   a. a
   b. b
   c. c
   d. d
   e. e

7. Refer to Figure 11-1. Which symbol represents a UJT?
   a. a
   b. b
   c. c
   d. d
   e. e

8. Refer to Figure 11-1. Identify the symbol for an SCS.
   a. a
   b. b
   c. c
   d. d
   e. e

9. Refer to Figure 11-1. Identify the diac symbol.
   a. a
   b. b
   c. c
   d. d
   e. e

10. Refer to Figure 11-1. What is the correct symbol for an SCR?
    a. a
    b. b
    c. c
    d. d
    e. e

11. You have a need to use a device to trigger an SCR. A good one to use might be
    a. an SCS.
    b. a UJT.
    c. a Schockley diode.
    d. a PUT.

12. Which of the following devices might best be used to control an ac motor?
    a. an SCS
    b. a PUT
    c. an SCR
    d. a diac

13. An SCR acts to control the speed of an ac motor by _____ the _____ of the pulse delivered to the motor.
    a. varying, width
    b. increasing, amplitude
    c. decreasing, gate width
    d. none of these

14. You need to design a circuit that will be a relaxation oscillator. The most likely device to use might be
    a. an SCR.
    b. a UJT.
    c. a triac.
    d. a Schockley diode.

15. You need a very efficient thyristor to control the speed of an ac fan motor. A good device to use would be
    a. a Schockley diode.
    b. a PUT.
    c. a triac.
    d. a BJT.

16. A device that might be used as a sensor to turn on a system of street lights is
    a. an SCR.
    b. a phototransistor.
    c. a solar cell.
    d. a thermistor.

17. You need a device to control a small DC motor directly when daylight occurs. The best choice is
    a. a photoresistor.
    b. a phototransistor.
    c. a photodiode.
    d. a LASCR.

18. You have the schematic diagram of several types of circuits. Which of these circuits most likely uses a triac?
    a. an oscillator
    b. an ac motor control
    c. a programmable oscillator
    d. an amplifier

19. You have a light-dimmer circuit using an SCR. In testing the circuit, you find that $I_G = 0$ mA and the light is still on. You conclude that the trouble might be one of the following:
    a. The SCR is open.
    b. The switch is faulty.
    c. The gate circuit is shorted.
    d. This is normal; nothing is wrong.

20. Your boss has asked you to recommend a thyristor that will enable you to turn it on with a pulse and also turn it off with a pulse. Which one of the following should you recommend?
    a. an SCR
    b. an SCS
    c. a PUT
    d. a triac

REVIEW OF KEY POINTS IN CHAPTER 12

OPERATIONAL AMPLIFIERS

INTRODUCTION TO OPERATIONAL AMPLIFIERS

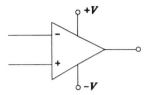

- Operational amplifiers or **op-amps** are **linear integrated circuits**.

- The op-amp has two inputs and one output.

- The two inputs are called **inverting** and **non-inverting.**

- On the op-amp symbol, the inverting input is marked (−) and the non-inverting (+).

- An **ideal amplifier** will have **infinite voltage gain, infinite input impedance, infinite bandwidth,** and **zero output impedance.**

- The op-amp comes close to having ideal amplifier characteristics.

THE DIFFERENTIAL AMPLIFIER

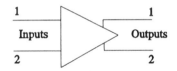

- Op-amp operation is based on a circuit known as a **differential amplifier**, which is found inside a typical op-amp.

- A differential amplifier has two inputs and two outputs.

- In **single-ended** input, one input is grounded and the other has the signal applied. Depending on the input to which the signal is applied, the output is either not-inverted or inverted.

- A signal applied to both inputs will produce an output voltage that is the instantaneous difference between the two inputs. This is the **differential input mode.**

- **Common-mode operation** involves applying two equal and in-phase signals.

- Two signals of this type cancel each other. This is called **common-mode rejection**. This is a very important aspect of op-amp applications.

- Common-mode signals are often **noise**. The ability of an op-amp to reject these in-phase noise signals is called the **common-mode rejection ratio (CMRR)**.

- **CMRR** = $A_{v(d)}/A_{cm}$ is the formula for expressing the CMRR. $A_{v(d)}$ is differential gain and A_{cm} is the common-mode gain.

- CMRR is usually expressed in dB. (dB)CMRR = 20 log $A_{v(d)}/A_{cm}$

OP-AMP PARAMETERS

- Many important **parameters or limitations** are found on an op-amp data sheet.

- A slight mismatch in the internal base-emitter voltages in the op-amp causes the output to have a small DC voltage even with no input voltages. This is **input offset voltage**. These voltages are small, about 2 mV or less.

- **Input bias current** is the total of the input currents required to operate the internal transistors. A typical value is 80 nA.

- Another small current is **input offset current**. This current is the difference between the input bias currents. This current is also very small, about 20 nA.

- **Open-loop voltage gain** is very high, typically about 200,000.

- The **CMRR** is expressed in dB and has typical values around 90 dB.

- The op-amp's ability to change the output voltage in step with a rapid input voltage change is called the **slew rate**. A typical slew rate is 0.5 V/μs.

- The op-amp can amplify DC voltages.

OP-AMPS WITH NEGATIVE FEEDBACK

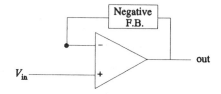

- **Negative feedback** (F.B.) takes a portion of the output voltage and feeds it back, out-of-phase with the input signal. This reduces the voltage gain and stabilizes the amplifier.

- A non-inverting amplifier has a voltage gain found by $A_v = 1 + R_f/R_i$.

- A **voltage follower amplifier** is a non-inverting amplifier with a closed loop between the output and the inverting input. The voltage gain is 1.

- An **inverting amplifier** has a voltage gain found by $A_v = -R_f/R_i$. The minus sign indicates a phase inversion.

- The op-amp input impedances are so high that the voltage difference between the two inputs is virtually 0 V. This is called **virtual ground**.

- The input impedance of a non-inverting amplifier is very high. It is not uncommon for Z_{in} to equal 20,000 MΩ.

- The output impedance of a non-inverting amplifier is very low, typically less than 1 Ω.

- An inverting amplifier has an input impedance equal to R_i. $\mathbf{Z_{in} = R_i}$

BIAS CURRENT AND OFFSET VOLTAGE COMPENSATION

- Bias current can be compensated for in a voltage follower amplifier by adding a series resistor, R_s, in the non-inverting input and a feedback resistor, R_f, in the feedback loop. $R_s = R_f$

- Other types of op-amp circuits can be bias current-compensated by adding a resistor, R_c, in the non-inverting input. The value should be $R_c = R_i \parallel R_f$.

- **Input offset voltage compensation** results in a small output voltage error in the range of μV to mV. Most op-amps have pins brought out to provide for external offset voltage compensation.

TECHNICAL TIPS

- The ideal amplifier would have an infinite voltage gain. With this gain, the amplifier could be used for any gain desired. Infinite input impedance means that the amplifier would never load a source because the input current would be zero. A zero output impedance would mean that any load current could be drawn from the amplifier with no restrictions as to magnitude. Of course, this ideal amplifier does not exist, but the op-amp does come reasonably close to these requirements.

- Most schematic diagrams of op-amp circuits do not show the $\pm V_{CC}$ supplies. This simplifies the diagram. Be sure to include these connections. Most op-amps are designed to operate with $\pm$ supplies. It is possible to use a single-ended supply, however.

- Virtual ground, where the voltage drop between inputs equals zero, is a most important attribute of an op-amp. This allows the op-amp to compare the voltages at the inputs, determine the difference, and multiply that difference by the gain. This is the basic operation of an op-amp.

- A valuable use for the op-amp is to reduce the common-mode noise that can exist at the inputs to an op-amp. A typical CMRR is 90 dB. This translates to the desired signal being amplified over 30,000 times more than the common-mode noise. A fairly large noise signal will be amplified so little as to be negligible.

- Op-amps are not generally power amplifiers. A typical op-amp can dissipate only about 50 mW. The extreme versatility of the op-amp allows its use in thousands of different types of circuit. If power is required to drive a loudspeaker, for example, some type of power amplifier would be required.

CHAPTER 12 QUIZ

Student Name _____ Date _____

1. A good op-amp has low voltage gain, low output impedance, and high input impedance.
   a. true
   b. false

2. The common-mode rejection ratio (CMRR) is a measure of an op-amp's ability to reject common-mode inputs.
   a. true
   b. false

3. An inverting amplifier has an input impedance equal to the feedback resistor R_f.
   a. true
   b. false

4. A non-inverting amplifier has a higher input impedance and a lower output impedance than the op-amp itself (without feedback).
   a. true
   b. false

5. All practical op-amps have input bias currents and voltages that produce output error voltages.
   a. true
   b. false

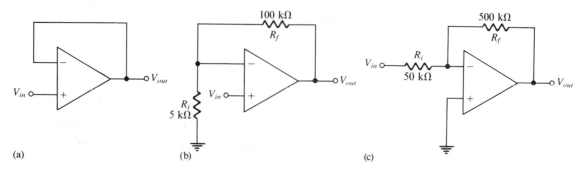

FIGURE 12-1

6. Refer to Figure 12-1(a). This amplifier is known as
   a. an inverting amplifier.
   b. a non-inverting amplifier.
   c. a voltage follower.
   d. a common-source amplifier.

7. Refer to Figure 12-1(b). This amplifier is known as
   a. an inverting amplifier.
   b. a non-inverting amplifier.
   c. a voltage follower.
   d. a common-source amplifier.

8. Refer to Figure 12-1(c). This amplifier is known as
   a. an inverting amplifier.
   b. a non-inverting amplifier.
   c. a voltage follower.
   d. a common-source amplifier.

9. Refer to Figure 12-1(a). A DC voltage of -1.2 V is applied. $V_{CC} = \pm 12$ V. What is the output voltage?
   a. 1.2 V
   b. -1.2 V
   c. 0 V
   d. 12 V

10. Refer to Figure 12-1(b). The voltage gain of this amplifier is
    a. 100.
    b. 5.
    c. 20.
    d. 21.

11. Refer to Figure 12-1(c). If an input signal of -0.5 V were applied, determine the output voltage.
    a. -5 V
    b. 5 V
    c. 10 V
    d. -10 V

100

12. Refer to Figure 12-1(c). The input impedance of this circuit is
    a. 500 kΩ.
    b. 10 kΩ.
    c. 50 kΩ.
    d. 5 kΩ.

13. Refer to Figure 12-1(c). You need an amplifier with an input impedance of 12 kΩ. You must **not** change the amplifier voltage gain. The new value of R_i would be _____ and the new value of R_f would be _____.
    a. 10 kΩ, 100 kΩ
    b. 13.3 kΩ, 120 kΩ
    c. 12 kΩ, 108 kΩ
    d. 12 kΩ, 120 kΩ

14. Refer to Figure 12-1(b). A DC input signal of −50 mV is applied. You would measure _____ from the inverting input to ground.
    a. 50 mV
    b. 1.05 V
    c. −1.05 V
    d. −50 mV

15. It takes an op-amp 22 µs to change its output from −15 V to +15 V. Determine the slew rate.
    a. 1.36 V/µs
    b. 0.68 V/µs
    c. −0.68 V/µs
    d. cannot determine

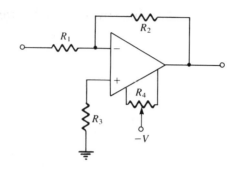

FIGURE 12-2

16. Refer to Figure 12-2. The purpose of R_1 and R_2 is
    a. for bias current compensation.
    b. for input offset voltage compensation.
    c. to set input impedance.
    d. to set input impedance and voltage gain.

17. Refer to Figure 12-2. The purpose of R_4 is
    a. for bias current compensation.
    b. for input offset voltage compensation.
    c. to set input impedance.
    d. to set input impedance and voltage gain.

18. Refer to Figure 12-2. The purpose of R_3 is
    a. for bias current compensation.
    b. for input offset voltage compensation.
    c. to set input impedance.
    d. to set input impedance and voltage gain.

19. Refer to Figure 12-2. If the value of R_1 decreases, the voltage gain will _____ and the input impedance will _____.
    a. increase, increase
    b. increase, decrease
    c. decrease, decrease
    d. decrease, increase

20. A voltage follower amplifier comes to you for service. You find the voltage gain to be 5.5 and the input impedance 22 kΩ. The probable fault in this amplifier, if any, is
    a. the gain is too low for this type of amplifier.
    b. the input impedance is too high for this amplifier.
    c. nothing is wrong. The trouble must be somewhere else.
    d. none of these.

REVIEW OF KEY POINTS IN CHAPTER 13

OP-AMP FREQUENCY RESPONSE, STABILITY, AND COMPENSATION

BASIC CONCEPTS AND OPEN-LOOP RESPONSE

- Op-amps will amplify DC as well as ac.

- **Open-loop gain** is the voltage gain without any feedback loop between the output and the input.

- Open-loop gain is quite large. It is sometimes called **large-signal voltage gain** on the data sheets.

- The frequency response of an open-loop amplifier is non-linear. The response falls off as the frequency is increased.

- The **3 dB open-loop bandwidth** is found by $BW = f_{c(high)}$.

- The open-loop gain for a particular frequency f, can be found by

$$A_{ol} = \frac{A_{ol(mid)}}{\sqrt{1 + \frac{f^2}{f_c^2}}}$$

Data sheets often refer to $A_{ol(mid)}$ as large signal voltage gain.

- When the open-loop gain is decreased by negative feedback, the bandwidth is increased.

CLOSED-LOOP RESPONSE

- An op-amp has a closed loop bandwidth that can be found by the formula
$BW_{cl} = BW_{ol}(1 + BA_{ol(mid)})$, where B = the attenuation in the feedback loop.

$$B = 1 + \frac{R_i}{R_f}$$

- A loop between the output and the inverting input will add negative feedback to the op-amp. The voltage gain can be set with external components. This is called **closed-loop gain**.

- **Unity-gain bandwidth** is the frequency where the gain falls off to unity or one.

POSITIVE FEEDBACK AND STABILITY

- Positive feedback occurs when the signal fed back from the output to the input is in phase with the input signal.

- The **phase margin**, θ_{pm}, is the amount of additional phase shift required to make the total phase shift around the loop 360°. $\theta_{pm} = 180° - |\theta_{tot}|$

- For mid-range frequencies, an op-amp must be operated at a closed loop gain with the roll-off rate not to exceed -20 dB/decade.

- Positive feedback applied to an op-amp will usually produce instability in the form of oscillations.

COMPENSATION

- Most op-amps have built-in **frequency compensation**.

- Frequency compensation is an internal or external RC network to provide stable operation at low gains.

- Op-amps with built-in frequency compensation have a relatively narrow BW. These op-amps are very stable, even at low gains.

- Many op-amps have terminals for the connection of external compensating capacitors. The designer can compromise between BW and gain with this type of device.

- Compensation increases the slew rate.

TECHNICAL TIPS

- The op-amp is a very good amplifier. Its characteristics, such as high gain, high input impedance, and low output impedance make this amplifier very valuable. A particularly good feature is the op-amp's ability to amplify DC voltages. Op-amps are used in many control applications where the signal is a DC voltage level. This is a large field for electronic technicians.

- There are some op-amps built especially for higher frequency operation. Most op-amps, however, have a limited high-frequency response. The gain-product bandwidth of typical op-amps may only be about 1 MHz. Your calculations in this chapter have shown these limitations.

CHAPTER 13 QUIZ

Student Name _____ Date _____

1. Open-loop gain of an op-amp is the voltage gain without feedback.
   a. true
   b. false

2. Closed-loop gain is the gain with negative feedback.
   a. true
   b. false

3. Open-loop gain is always smaller than the closed-loop gain.
   a. true
   b. false

4. Negative feedback lowers the bandwidth and increases the voltage gain.
   a. true
   b. false

5. Compensation reduces bandwidth and increases the slew rate.
   a. true
   b. false

6. An op-amp has a midrange gain of 100,000 and a cutoff frequency of 400 Hz. Find the open-loop gain at a frequency of 300 Hz.
   a. 800
   b. 8000
   c. 80000
   d. 800000

7. An op-amp has a midrange gain of 75,000 and a cutoff frequency of 100 Hz. Find the open-loop gain at 1 kHz.
   a. 7987
   b. 7463
   c. 7124
   d. 7002

8. An RC network has R = 47 kΩ and C = 0.22 μf. What is the cutoff frequency?
   a. 154 Hz
   b. 1540 Hz
   c. 1.54 Hz
   d. 15.4 Hz

9. An RC network has R = 500 kΩ and C = 10 pF. Find the value of f_c.
   a. 31847 Hz
   b. 31.847 kHz
   c. 0.031847 MHz
   d. all of the above
   e. none of the above

10. A certain op-amp has an open-loop voltage gain of 150,000. What is this gain expressed in dB?
    a. 51.7 dB
    b. 103.5 dB
    c. 150,000 dB
    d. 5.18 dB

11. The midrange open-loop gain of an op-amp is 135 dB. With negative feedback this gain is reduced to 72 dB. The closed-loop gain is
    a. 135 dB.
    b. 72 dB.
    c. 207 dB.
    d. 63 dB.

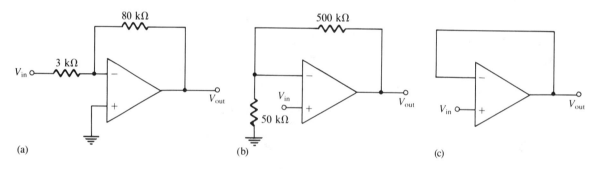

FIGURE 13-1

12. Refer to Figure 13-1(a). Find the midrange gain of this amplifier.
    a. 26.7
    b. −26.7
    c. 27.7
    d. −27.7

13. Refer to Figure 13-1(b). Find the midrange gain of this op-amp amplifier.
    a. 10
    b. 11
    c. −10
    d. −11

14. Refer to Figure 13-1(c). The midrange voltage gain of this amplifier is
    a. 0.5.
    b. 27.7.
    c. −11.
    d. 1.

15. Refer to Figure 13-1(a). The op-amp has a unity-gain bandwidth of 3 MHz. Determine the BW of the circuit.
    a. 3 MHz
    b. 30 kHz
    c. 108.4 kHz
    d. infinite in width

16. Refer to Figure 13-1(b). The op-amp has a unity-gain bandwidth of 1.7 MHz. Find the bandwidth of the circuit.
    a. 155 MHz
    b. 155 kHz
    c. 155 Hz
    d. 15.5 Hz

17. Refer to Figure 13-1(c). The unity-gain bandwidth of this op-amp is 10.4 kHz. What is the bandwidth of the circuit?
    a. 10.4 kHz
    b. 15.5 kHz
    c. 3 MHz
    d. 16.7 kHz

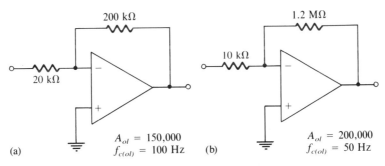

FIGURE 13-2

18. Refer to Figure 13-2(a). Determine the bandwidth.
    a. 1 MHz
    b. 1.5 MHz
    c. 1 kHz
    d. 1.5 kHz

19. Refer to Figure 13-2(b). Calculate the bandwidth.
    a. 8.33 MHz
    b. 833 kHz
    c. 83.3 kHz
    d. 8.33 kHz

20. Negative feedback added to an op-amp _____ the bandwidth and _____ the gain.
    a. increases, increases
    b. increases, decreases
    c. decreases, decreases
    d. decreases, increases

REVIEW OF KEY POINTS IN CHAPTER 14
BASIC OP-AMP CIRCUITS

COMPARATORS

- Op-amps are often used to compare the amplitude of two voltages. This circuit arrangement is called a **comparator**.

- A comparator has a reference voltage on one input and a signal voltage on the other.

- The magnitude and polarity of the two voltages will cause the output to saturate in either the positive or negative direction.

- The high open-loop gain will cause even a very small voltage difference between the inputs to cause saturation in the output.

- If a sine wave is applied to one input and the other is grounded, the output will change state whenever the input voltage crosses zero volts. This is called a **zero-level detector**.

- A different reference voltage can be applied, and the output will change state from positive to negative at the reference voltage. The output will also change state from negative to positive. This application is called a **non-zero-level detector**.

- A voltage divider can be used to supply the reference voltage.

- Zener diodes are also used for setting the reference voltage.

- The state can change on positive- or negative-going signal voltages, depending on the input used.

- Noise, along with the sine wave signal, can affect when the signal crosses the reference voltage. This will cause false changes in state.

- **Hysteresis** can be used to reduce the false triggering of the op-amp due to noise or other glitches.

- Hysteresis is the condition in which the change from positive to negative is at a different level than the change from negative to positive.

- An op-amp with this kind of circuit is called a **Schmitt trigger**.

- The output voltage can be limited to voltages other than saturation by use of a zener diode in the feedback loop. This is called **output bounding**.

- Comparators are used extensively in industrial control systems for many types of sensing circuits.

SUMMING AMPLIFIERS

- An op-amp circuit that can be used to sum or add voltages is called a **summing amplifier**. This is sometimes called an **adder**.

- If the input resistors and the feedback resistor are equal, $R_1 = R_2 = R_f$, then the output voltage is equal to the sum of the input voltages.
$$V_{OUT} = -(V_{IN_1} + V_{IN_2} + \cdots + V_{IN_n})$$

- Gain can be added to the summing amplifier by making R_f larger than the input resistors. The formula for the output voltage becomes
$$V_{OUT} = -R_f/R(V_{IN_1} + V_{IN_2} + \cdots + V_{IN_n})$$

- An **averaging amplifier** can be made by setting R_f/R equal to $1/n$, where n = number of inputs.

- A different weight can be given to each input by varying the value of each R_{IN}. This circuit then becomes a **scaling adder**. The output is found by this formula:
$$V_{OUT} = -((R_f/R_1)V_{IN_1} + (R_f/R_2)V_{IN_2} + \cdots + (R_f/R_n)V_{IN_n})$$

THE INTEGRATOR AND DIFFERENTIATOR

- An **integrator** circuit can be made by replacing the feedback resistor with a capacitor.

- The output will be a negative ramp when a positive input voltage is applied.

- The output voltage will ramp negative until $-V_{SAT}$ is reached.

- A square wave input will produce a triangle wave output.

- A **differentiator circuit** is made by reversing the positions of the resistor and capacitor in the integrator.

- The circuit will differentiate the incoming signal in the output.

- Differentiation is the opposite of integration.

- A differentiator with a triangle wave input will produce a square wave output.

TROUBLESHOOTING

- Op-amps do fail. However, since an op-amp is an integrated circuit, the failure rate of the typical op-amp is low.

- Other components in the circuit, such as capacitors, diodes, or resistors, will cause circuit problems more often than the op-amp.

TECHNICAL TIPS

- Comparators are an often-used application for op-amps. Remember that the open-loop voltage gain is very high. Almost any input signal will cause the op-amp output to saturate at a voltage that is called V_{sat}. This voltage can be either positive or negative. There are many transistors in a typical op-amp, and each of these must be biased correctly. These bias voltages drop about 2 V. If the op-amp is saturated, V_{sat} is approximately 2 V less than V_{CC}. For example, if $V_{CC} = 15$ V, then $V_{sat} \cong 13$ V.

- There is an easy way to determine the output direction of saturation of a comparator. Select one input as the reference. Compare the voltage on the other input to the reference: is it positive or negative with respect to the reference? The output will be either $\pm V_{sat}$, depending on the reference selected.

- Let's try one. You have -6 V on the inverting input and 2 V on the non-inverting input. Select the non-inverting input as your reference. The -6 V is negative compared to the reference voltage of 2 V. A negative voltage on the inverting input will produce a positive V_{sat} output. If you select the inverting input of -6 V as the reference, the positive 2 V on the non-inverting input will be positive compared to the reference. The output will be $+V_{sat}$. Either way, you get the same result.

- A summing amplifier depends on virtual ground for successful operation. This virtual ground isolates the currents in each input from one another. There are no currents flowing from one input into the other. The summing amplifier will also produce the instantaneous sum of ac signals on the inputs. A DC level can be added to an ac signal as well.

- A very simple audio mixer circuit can be made from a scaling adder. If a guitar is connected to input one, a bass guitar connected to input two, and a voice microphone connected to input three, then the circuit will mix them and give the desired output. Change each R_i to a potentiometer, and you have individual gain controls on each channel. If a master gain control is desired, change R_f to a potentiometer as well. Of course, this circuit would need some refinements to provide input impedances to match the input microphones. Another op-amp could be used in each channel input for this purpose. The possibilities are endless.

- The op-amp integrator offers advantages over the integrator built from discrete components. The capacitor in an RC integrator charges in a non-linear fashion. It is difficult to get a linear output. There will always be some curve to the output ramp. The op-amp integrator will produce a linear ramp output voltage. This is due to the charging current being limited to a fixed value by Ohm's Law.

- A good tip to remember is an inverting amplifier has its output voltage equal to the voltage across the feedback resistor. $V_{out} = V_{R_f}$

CHAPTER 14 QUIZ

Student Name _____ Date _____

1. In an op-amp comparator, when the input voltage exceeds a reference voltage, the output changes state.
   a. true
   b. false

2. Bounding allows the output of a comparator to be an unlimited voltage.
   a. true
   b. false

3. A positive feedback network for hysteresis improves an op-amp comparator's noise immunity.
   a. true
   b. false

4. Operational amplifiers are never used as nonlinear devices.
   a. true
   b. false

5. The output voltage of a summing amplifier is proportional to the sum of the input voltages.
   a. true
   b. false

6. An op-amp has an open-loop gain of 90,000. $V_{sat} = \pm 13$ V. A differential voltage of 0.1 V_{p-p} is applied between the inputs. What is the output voltage?
   a. 13 V
   b. -13 V
   c. 13 V_{p-p}
   d. 26 V_{p-p}

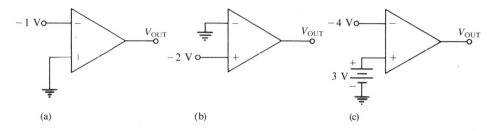

FIGURE 14-1

7. Refer to Figure 14-1(a). Determine the output voltage.
   a. 1 V
   b. −1 V
   c. $+V_{sat}$
   d. $-V_{sat}$

8. Refer to Figure 14-2(b). What is the output voltage?
   a. 2 V
   b. −2 V
   c. $+V_{sat}$
   d. $-V_{sat}$

9. Refer to Figure 14-1(c). With the inputs shown, determine the output voltage.
   a. 7 V
   b. −7 V
   c. $+V_{sat}$
   d. $-V_{sat}$

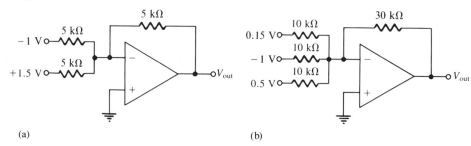

FIGURE 14-2

10. Refer to Figure 14-2(a). What is the output voltage?
    a. 0.5 V
    b. −0.5 V
    c. 2 V
    d. −2 V

11. Refer to Figure 14-2(b). Determine the output voltage, V_{OUT}.
    a. 1.05 V
    b. −0.35 V
    c. 0.35 V
    d. −1.05 V

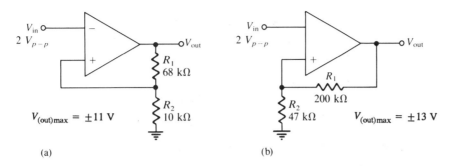

FIGURE 14-3

12. Refer to Figure 14-3(a). Determine the upper trigger point.
    a. $V_{(out)max}$
    b. $-V_{(out)max}$
    c. -1.41 V
    d. $+1.41$ V

13. Refer to Figure 14-3(b). Determine the lower trigger point.
    a. $+V_{(out)max}$
    b. $-V_{(out)max}$
    c. -2.47 V
    d. $+2.47$ V

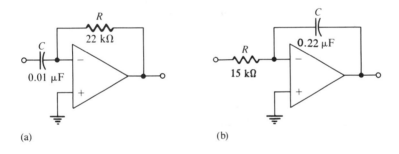

FIGURE 14-4

14. Refer to Figure 14-4(a). This circuit is known as
    a. a non-inverting amplifier.
    b. a differentiator.
    c. an integrator.
    d. a summing amplifier.

15. Refer to Figure 14-4(b). This circuit is known as
    a. a non-inverting amplifier.
    b. a differentiator.
    c. an integrator.
    d. a summing amplifier.

16. Refer to Figure 14-4(b). A square-wave input is applied to this amplifier. The output voltage is most likely to be
    a. a square wave.
    b. a triangle wave.
    c. a sine wave.
    d. no output.

17. Refer to Figure 14-4(b). If V_{in} = 5 V, the rate of change of the output voltage in response to a single pulse input is:
    a. 15.2 mV/μs
    b. 1.52 V/μs
    c. 1.52 mV/μs
    d. 15.2 mV/μs

18. The output of a Schmitt trigger is a
    a. square wave.
    b. sawtooth wave.
    c. since wave.
    d. triangle wave.

19. A Schmitt trigger is
    a. a comparator with only one trigger point.
    b. a comparator with hysteresis.
    c. a comparator with three trigger points.
    d. none of the above.

20. An integrator circuit
    a. uses a resistor in its feedback circuit.
    b. uses an inductor in its feedback circuit.
    c. uses a capacitor in its feedback circuit.
    d. a or c

REVIEW OF KEY POINTS IN CHAPTER 15

MORE OP-AMP CIRCUITS

INSTRUMENTATION AMPLIFIERS

- There are many applications for an amplifier to be used at the end of a long cable. This cable may have a very large common-mode signal due to various types of noise induced in an industrial setting. An amplifier built for this application is called an **instrumentation amplifier**.

- The typical characteristics are high input impedance, Z_{in}, high voltage gain, A_v, and high **CMRR**. The CMRR may be as high as 100 dB. The higher the CMRR, the better the amplifier.

- An instrumentation amplifier consists of three op-amps in one IC chip and seven resistors.

- An instrumentation amplifier voltage gain is set by a single external resistor. The value of this resistor, R_G, can be found using the expression $R_G = 2R/(A_{Cl} - 1)$, where $R = R_1 = R_2$.

- A typical application of an instrumentation amplifier is where small signals are embedded in large common-mode noise.

ISOLATION AMPLIFIERS

- An **isolation amplifier** consists of three electrically isolated sections: **input, output,** and **power**.

- For isolation purposes, most isolation amplifiers use transformer coupling.

- An isolation amplifier provides DC isolation between input and output.

- Basic applications of an isolation amplifier are: medical instrumentation, power plant instrumentation, industrial processing, and automated testing.

- The input voltage gain of an isolation amplifier can be found using:
$A_{v(in)} = 1 + R_F/R_1$.

- The output voltage gain of an isolation amplifier can be found using:
$A_{v(output)} = (75 \text{ k}\Omega + R_{ext})/30 \text{ k}\Omega$.

OPERATION TRANSCONDUCTANCE AMPLIFIERS (OTAS)

- The **OTA** is primarily a voltage-to-current amplifier.

- The **output current** of an OTA is the **product** of the input voltage and the transconductance. Mathematically, $I_{out} = g_m \times V_{in}$, where g_m is the transconductance.

- The **transconductance** of an OTA varies with the bias current which causes the gain to vary with the bias voltage.

- The **DC voltage** of an OTA should not exceed ± 15 V.

- As the bias current increases, both the input and the output resistances of an OTA decrease.

LOG AND ANTILOG AMPLIFIERS

- An amplifier that produces an output that is proportional to the logarithm of an input is known as a **logarithmic amplifier**.

- An amplifier that takes the inverse log of an input is known as an **antilogarithmic amplifier**.

- A log amplifier uses a **BJT** in its feedback loop.

- An antilog amplifier uses a **BJT in series** with the input.

- A basic application of logarithmic amplifier is analog multiplication and division.

CONVERTERS AND OTHER OP-AMP CIRCUITS

- An amplifier that provides a constant load current with any change in the load resistance is known as a **constant current source**.

- The load current can be found mathematically using $I_L = V_{in}/R_i$.

- Another application of an op-amp is the **current-to-voltage converter**. This type of circuit will provide an output voltage that is proportional to the input current. That is, as the input current changes, the output voltage will change proportionally.

- The output voltage can be found mathematically using $V_{out} = I_i R_f$.

- A **voltage-to-current** converter is a circuit that is used to control an output current with a varying input voltage.

- The output current can be found mathematically using $I_L = V_{in}/R_1$.

- The addition of a diode and a capacitor at the output will produce a circuit that will act as a **peak detector**.

- The purpose of the capacitor is to store the detected input peak voltage for a limited amount of time.

TECHNICAL TIPS

- Any time there is a safety hazard related to electrical equipment, it is important to use an isolation amplifier.

- It is important to remember that the third op-amp of an instrumentation amplifier is used as a unity-gain differential amplifier.

- If the output voltage of an instrumentation amplifier decreases, check the gain of the amplifier by checking the external resistance R_G.

- An OTA can be used as an amplitude modulator.

- To drive a device into its saturation state, an OTA can be used in a Schmitt-trigger configuration. Recall that a Schmitt-trigger is a comparator with hysteresis.

CHAPTER 15 QUIZ

Student Name _____ Date _____

1. One of the key characteristics of an instrumentation amplifier is high input impedance.
   a. true
   b. false

2. To construct an instrumentation amplifier, two op-amps and seven resistors are needed.
   a. true
   b. false

3. An isolation amplifier provides ac isolation between input and output.
   a. true
   b. false

4. One of the principal areas of application for an isolation amplifier is power plan instrumentation.
   a. true
   b. false

5. The main purpose of an instrumentation amplifier is to amplify small signals riding on large common-mode voltages.
   a. true
   b. false

6. Instrumentation amplifiers are commonly used in environments with low common-mode noise
   a. true
   b. false

7. In an isolation amplifier, the third op-amp is used as a unity-gain differential amplifier.
   a. true
   b. false

8. An OTA is primarily a current-to-voltage amplifier.
   a. true
   b. false

9. OTA stands for Operational Transistor Amplifier.
   a. true
   b. false

10. A log amplifier has a BJT in the feedback loop.
    a. true
    b. false

11. An antilog amplifier has a BJT in series with the input.
    a. true
    b. false

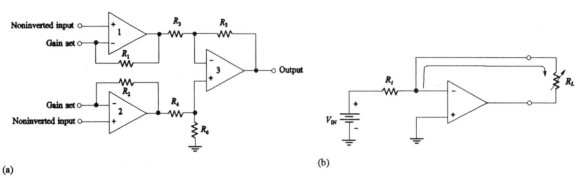

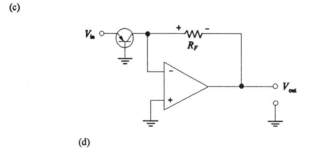

FIGURE 15-1

12. Refer to Figure 15-1(a). This circuit is a setup for
    a. an antilog amplifier.
    b. a constant-current source.
    c. an instrumentation amplifier.
    d. an isolation amplifier.

13. Refer to Figure 15-1(b). This circuit is a setup for
    a. an antilog amplifier.
    b. a constant-current source.
    c. an instrumentation amplifier.
    d. an isolation amplifier.

14. Refer to Figure 15-1(c). This circuit is a setup for
    a. an antilog amplifier.
    b. a constant-current source.
    c. an instrumentation amplifier.
    d. an isolation amplifier.

15. Refer to Figure 15-1(d). This circuit is a setup for
    a. an antilog amplifier.
    b. a constant-current source.
    c. an instrumentation amplifier.
    d. an isolation amplifier.

16. The OTA has a _____ input impedance and a _____ CMRR.
    a. high, low
    b. low, high
    c. high, high
    d. low, high

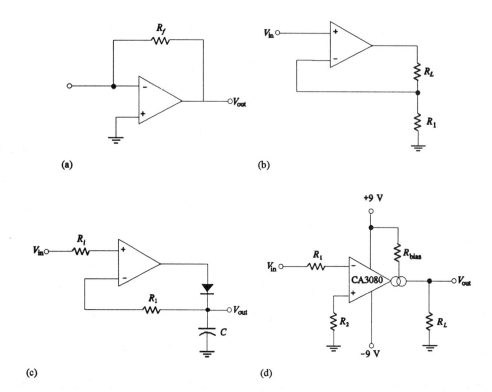

FIGURE 15-2

17. Refer to Figure 15-2. Which circuit is known as a voltage-to-current converter?
    a. a
    b. b
    c. c
    d. d

18. Refer to Figure 15-2. Which circuit is known as a current-to-voltage converter?
    a. a
    b. b
    c. c
    d. d

19. Refer to Figure 15-2. Which circuit is known as an OTA?
    a. a
    b. b
    c. c
    d. d

20. Refer to Figure 15-2. Which circuit is known as a peak detector?
    a. a
    b. b
    c. c
    d. d

REVIEW OF KEY POINTS IN CHAPTER 16

ACTIVE FILTERS

BASIC FILTER RESPONSES

- A **filter** is a circuit that passes certain frequencies easily, and attenuates all other frequencies.

- An **active filter** is a circuit that includes an RC filter network followed by an op-amp to provide gain and impedance characteristics.

- A **low-pass filter** passes low frequencies from DC up to the cutoff frequency f_c. At f_c the response is down 3 dB or 0.707 from the response in the **band-pass**.

- The passband of a filter is that group of frequencies easily passed.

- The bandwidth of a low-pass filter is equal to f_c. **BW = f_c**

- A **high-pass filter** rejects all frequencies below f_c, and passes all frequencies above f_c.

- At the cutoff frequency for both high-pass and low-pass filters, $X_C = R$ and $f_c = 1/2\pi RC$.

- A **roll-off rate** of -20 dB/decade occurs for a **single-pole filter stage**.

- A single-pole stage consists of one resistor and one capacitor.

- A **band-pass filter** passes all frequencies between an upper and lower f_c. All other frequencies above or below these frequencies are attenuated.

- The bandwidth is found by **BW = $f_{c2} - f_{c1}$**.

- The **center frequency (f_o)** is the mean of the two cutoff frequencies and is found by

$$f_o = \sqrt{f_{c1}f_{c2}}$$

- The **bandwidth** of a band-pass filter is **BW = f_o/Q**.

- A **band-stop filter** rejects all frequencies between two cutoff frequencies and passes all others.

- Other popular names for a band-stop filter are **notch filter, band-reject filter,** and **band-elimination filter.**

FILTER RESPONSE CHARACTERISTICS

- A very common active filter is the **Butterworth**.

- A Butterworth filter provides a very flat response in the passband and a roll-off rate of 20 dB/decade.

- The Butterworth filter does have a phase shift, and for this reason it is not often used in pulse applications. Overshoot will occur with pulses applied.

- Another filter, the **Chebyshev**, has the characteristic of a very rapid roll-off rate greater than 20 dB/decade/pole. This filter will have ripples in the response within the passband.

- The **Bessel filter** is used for filtering pulse waveforms. There is little phase shift, so pulse waveforms are not distorted.

- The **damping factor** determines the characteristics that a filter will have. $DF = 2 - R_1/R_2$, where R_1/R_2 is the feedback resistor ratio in the negative feedback loop. This damping factor must have a definite value to determine the type of filter characteristic desired, such as a Butterworth or a Bessel. The number of filter poles also must be considered in the value of the damping factor.

- The cutoff frequency of an active filter is determined by the values of R and C.

- A **first-order (single-pole)** filter has $f_c = 1/2\pi RC$.

- The number of poles determines the roll-off rate. Each pole adds a roll-off rate of 20 dB/decade.

ACTIVE LOW-PASS FILTERS

- An active low-pass filter consists of an RC filter network followed by an op-amp amplifier.

- The RC filter network output is connected to the non-inverting input of the op-amp. Remember the voltage gain $A_v = R_1/R_2 + 1$.

- A **Sallen and Key low-pass filter** is a two-pole variety. If the filter resistors are equal and the capacitors are equal, the cutoff frequency is found by $f_c = 1/2\pi RC$.

- The damping factor of 1.414 required to obtain a second-order Butterworth filter will require an R_1/R_2 ratio of 0.586.

- Low-pass filters can be **cascaded** to obtain a greater roll-off rate. For example, a two-pole filter followed by another two-pole filter will produce a roll-off rate of 80 dB/decade.

ACTIVE HIGH-PASS FILTERS

- A high-pass active filter follows the same theory as a low-pass filter. The positions of the resistors and capacitors are reversed.

- High-pass filters can be cascaded in a similar manner to low-pass filters.

ACTIVE BAND-PASS FILTERS

- An active band-pass filter can be made using cascaded high-pass and low-pass filters.

- Another band-pass filter is the **multiple-feedback filter**. This filter uses a high-pass and a low-pass section providing two feedback paths.

- A **state-variable filter** is often used as a band-pass filter.

- State-variable filters use a summing amplifier and two integrators to provide the band-pass output.

ACTIVE BAND-STOP FILTERS

- Band-stop filters can use a state-variable filter or a multiple-feedback filter.

TECHNICAL TIPS

- Recall your previous studies on resonant circuits. At resonance the impedance of a parallel resonant circuit was maximum. The bandwidth was dependent on the quality factor, Q. Here we are dealing not with resonant circuits directly, but with circuits that can give a frequency response similar to that of resonant LC circuits. A narrow band filter is a band-pass filter with a Q greater than 10. A wide band filter has a Q of less than 10. Remember, as Q increases, the bandwidth narrows.

- Band-stop filters occupy an important place in electronic applications. Communication circuits found in short-wave radios, for example, use band-stop filters to reject another signal close in frequency to the desired signal. These close frequencies are unwanted. Most of these filters are tunable with a variable component such as a potentiometer. The user can then vary the f_r of the filter to drop the unwanted frequency into a hole, so to speak. In other words, the filter rejects the signal.

- The Butterworth type of filter is most commonly used. The flat frequency response is the main advantage to this filter. It is also very easy to design and construct. The low cost of most op-amps and of the resistors and capacitors make this type of filter desirable.

- The frequency response curve is widely used in textbooks, manufacturers' data sheets, and in consumer literature. Many students expect to see a typical low-pass or high-pass response curve in the lab. Some relatively special equipment is required to see the curve on an oscilloscope screen. The curve is merely a plot of the output voltage as the frequency is increased. The circuit should be set up using a dual-trace oscilloscope, with one trace for the input signal and the other for the output. As you vary the input frequency, be sure to watch the input trace and keep the amplitude constant. Your table of output data is then plotted on graph paper. The resulting curve is the frequency response curve for the filter under test.

- A band-pass filter consists of a high-pass and a low-pass filter section in some configuration. If a wide bandwidth is desired, remember that the f_c of the low-pass section is the higher cutoff frequency in value. The opposite is true for the lowest f_c. Be sure your RC values reflect this. The f_{cl} is from the high-pass filter section and the f_{ch} is from the low-pass filter section.

CHAPTER 16 QUIZ

Student Name _____ Date _____

1. The bandwidth of a band-pass filter is the sum of the two cutoff frequencies.
   a. true
   b. false

2. Butterworth filters have the characteristic of a very flat response in the band-pass and a roll-off of 20 dB/decade.
   a. true
   b. false

3. Filters with Bessel characteristics are used for filtering pulse waveforms.
   a. true
   b. false

4. In filters, a single RC network is called a pole.
   a. true
   b. false

5. A band-pass filter passes all frequencies within a band between a lower and upper cutoff frequency.
   a. true
   b. false

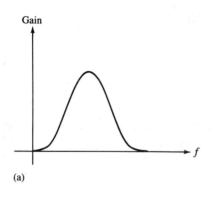
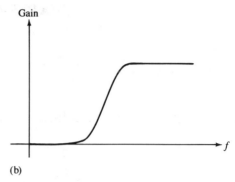
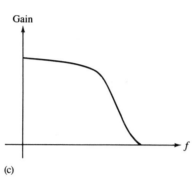
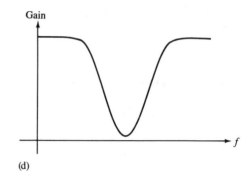

FIGURE 16-1

6. Refer to Figure 16-1. Identify the frequency response curve for a band-reject filter.
   a. a
   b. b
   c. c
   d. d

7. Refer to Figure 16-1. Identify the frequency response curve for a low-pass filter.
   a. a
   b. b
   c. c
   d. d

8. Refer to Figure 16-1. Identify the frequency response curve for a high-pass filter.
   a. a
   b. b
   c. c
   d. d

9. Refer to Figure 16-1. Identify the frequency response curve for a band-pass filter.
   a. a
   b. b
   c. c
   d. d

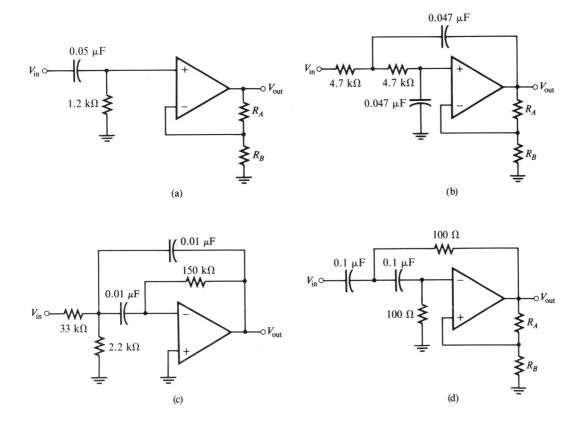

FIGURE 16-2

10. Refer to Figure 16-2(a). This is a _____ filter, and it has a cutoff frequency of _____.
    a. high-pass, 21 Hz
    b. low-pass, 21 Hz
    c. high-pass, 2.65 kHz
    d. low-pass, 2.65 kHz

11. Refer to Figure 16-2(b). The cutoff frequency of this filter is _____, and the circuit is known as a _____.
    a. 721 Hz, low-pass filter
    b. 721 Hz, high-pass filter
    c. 72 Hz, low-pass filter
    d. 721 Hz, band-pass filter

12. Refer to Figure 16-2(c). This is a _____ filter.
    a. band-pass
    b. band-stop
    c. high-pass
    d. low-pass

13. Refer to Figure 16-2(d). This circuit is known as a _____ filter, and the f_c is _____.
    a. high-pass, 1.59 kHz
    b. band-pass, 15.9 kHz
    c. low-pass, 15.9 kHz
    d. high-pass, 15.9 kHz

14. Refer to Figure 16-2(b). $R_A = 2.2$ kΩ and $R_B = 1.2$ kΩ. This filter is probably a
    a. Butterworth type.
    b. Bessel type.
    c. Chebyshev type.

15. Refer to Figure 16-2(c). The roll-off of this filter is about
    a. 20 dB/decade.
    b. 40 dB/decade.
    c. 60 dB/decade.
    d. 80 dB/decade.

16. Refer to Figure 16-2(a). The roll-off of the circuit shown is about
    a. 20 dB/decade.
    b. 40 dB/decade.
    c. 60 dB/decade.
    d. 80 dB/decade.

17. A low-pass filter has a cutoff frequency of 1.23 kHz. Determine the bandwidth of the filter.
    a. 2.46 kHz
    b. 1.23 kHz
    c. 644 Hz
    d. not enough information given

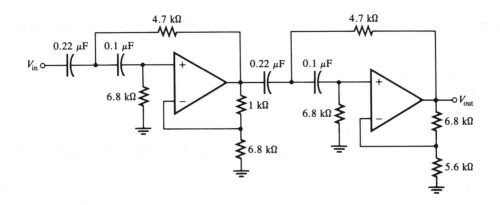

FIGURE 16-3

18. Refer to Figure 16-3. This is a _____ filter.
    a. low-pass
    b. high-pass
    c. band-pass
    d. band-stop

19. Refer to Figure 16-3. This filter has a roll-off rate of
    a. 20 dB/decade.
    b. 40 dB/decade.
    c. 60 dB/decade.
    d. 80 dB/decade.

20. Refer to Figure 16-3. Increasing the values of the filter section resistors in this circuit will cause the f_c to
    a. increase.
    b. decrease.
    c. remain the same.
    d. increase and then decrease.

REVIEW OF KEY POINTS IN CHAPTER 17
OSCILLATORS AND THE PHASE-LOCKED LOOP

THE OSCILLATOR

- A circuit that produces a repeating waveform with only DC as an input is called an **oscillator**.

- Oscillators are used in many electronic systems, such as radio, TV, telephones, and industrial systems.

- An oscillator can produce many types of output waveforms, such as sine waves, triangle waves, or square waves.

OSCILLATOR PRINCIPLES

- Oscillators operate with positive feedback.

- The output of an oscillator is fed back to the input **in-phase** through some type of feedback network.

- The in-phase feedback is called **positive feedback**.

- The active element in an oscillator is an amplifier. The amplifier may be of several varieties, such as BJT, FET, or op-amp.

- The voltage gain of the amplifier must be enough to overcome the loss in the feedback network.

- To start oscillation, it is necessary to have a voltage gain greater than 1 so that the waveform output will build up in amplitude.

- The voltage gain must be reduced to about unity to sustain the oscillations after they have started, and the phase shift around the feedback loop must be 0°.

OSCILLATORS WITH RC FEEDBACK CIRCUITS

- The phase-shifting feedback network of an oscillator can consist of an RC network.

- A very common RC feedback oscillator is the **Wien-bridge oscillator.**

- A Wien-bridge oscillator uses a **lead-lag network** in the feedback loop.

- A lead-lag network changes the phase relationship from a leading to a lagging phase angle as the frequency changes.

- 0° phase shift is achieved at a frequency called the **resonant frequency**.

- In a Wien-bridge oscillator, only the desired resonant frequency is fed back in-phase. The output is a sine wave at the resonant frequency f_r. $f_r = 1/2\pi RC$ where $R = R_1 = R_2$ and $C = C_1 = C_2$.

- Wien-bridge oscillators often use back-to-back zener diodes in parallel with the feedback resistor in the negative loop. These diodes increase the gain required during the start-up phase of the oscillator.

- A **phase-shift oscillator** uses a three-section RC network to provide the necessary in-phase feedback.

- The resistors and capacitors in the RC network usually have the same value, R and C. The frequency of oscillation is found by $f_r = 1/2\pi\sqrt{6}\,RC$.

- A **twin-T oscillator** uses two RC networks. One is a T-type low-pass filter and the other is a T-type high-pass filter. These filters are in the negative feedback loop. These filters act as a band-reject filter. The only frequency where negative feedback is minimum is the resonant frequency.

- Most oscillators that use RC feedback networks are usable at frequencies up to about 1 MHz.

OSCILLATORS WITH LC FEEDBACK CIRCUITS

- High frequency oscillators are frequently of the LC feedback type.

- LC oscillators use a **parallel resonant or tank** circuit to establish the resonant frequency.

- The phase shift across a tank circuit is 180°.

- A **Colpitts** oscillator uses a tank circuit with two capacitors in series. The feedback connection is from the junction between the two capacitors.

- The resonant frequency, f_r, of a Colpitts oscillator is found by $f_r = \dfrac{1}{2\pi\sqrt{LC}}$, where C is the total series capacitance in the tank.

- A load applied to the oscillator will act to change the resonant frequency by reducing the Q.

- A variation of the Colpitts is the **Clapp oscillator**.

- The Clapp oscillator uses the two series capacitors, as does the Colpitts. In addition, it uses a third capacitor in series with the inductor in the tank. The frequency is determined largely by the value of this extra capacitor.

- An oscillator similar to the Colpitts is the **Hartley oscillator**. This circuit uses a split inductor instead of the two capacitors.

- The resonant frequency for the Hartley oscillator is found with the familiar formula mentioned above.

- The **Armstrong oscillator** uses a feedback loop that is transformer-coupled from the tank circuit back to the input.

- A very stable and accurate oscillator is the **crystal oscillator**.

- Crystal oscillators usually incorporate a **quartz crystal**. These crystals exhibit what is called the **piezoelectric effect**.

- This piezoelectric effect causes a slice of quartz crystal to vibrate at a natural frequency when a mechanical force is applied to it. While vibrating, a voltage will be produced across the crystal at its natural frequency.

- Conversely, when an ac voltage is applied across the crystal, it will mechanically vibrate at its natural frequency.

- A crystal acts as a mechanical tank circuit. The frequency of oscillation is dependent upon the size and shape of the crystal.

- As in a tank circuit, a crystal in series with the feedback will provide minimum impedance at f_r. A crystal in parallel with the feedback loop will provide maximum impedance at the f_r.

NONSINUSOIDAL OSCILLATORS

- An oscillator can produce **triangle waves**. A simple circuit uses a comparator followed by an integrator.

- A **voltage-controlled oscillator (VCO)** is an oscillator whose frequency can be changed by an applied voltage.

THE 555 TIMER AS AN OSCILLATOR

- A **multivibrator** is a circuit that produces square waves.

- There are two general types of multivibrators, **astable or free-running**, and **monostable or one-shot**. In addition, there is a bistable multivibrator used in digital circuits.

- An astable multivibrator produces a constant train of square waves as long as the circuit is turned on.

- A monostable multivibrator produces one output square wave for each input trigger.

- The 555 timer consists of two comparators, a flip-flop, a discharge transistor, and a buffer output all mounted in an IC chip.

- To produce the astable state, the 555 timer uses an external capacitor and two resistors. The frequency of oscillation is found by $f = 1.44/(R_1 + 2R_2)C_{ext}$.

- The duty cycle of the timer in the astable state can be found by
  Duty cycle = $(R_1 + R_2)/(R_1 + 2R_2) \times 100\%$.

- The capacitor charges through R_1 and R_2 and discharges through R_2. A minimum duty cycle of 50% can be achieved. A lower duty cycle requires a diode in parallel with R_2. Under this condition, **Duty cycle = $R_1/(R_1 + R_2) \times 100\%$**.

THE PHASE-LOCKED LOOP

- A **phase-locked loop (PLL)** is an electronic feedback circuit. It acts to lock onto a frequency and remain at that frequency as long as a certain phase relationship exists.

- The circuit consists of a **phase detector, a low-pass filter,** and a **VCO**.

- Phase-locked loops are used extensively in all types of communications equipment, including TV receivers, FM radios, and amateur radio receivers.

TECHNICAL TIPS

- An oscillator is basically an amplifier with its own input. The feedback must be in-phase, or positive. The gain must be just enough to overcome the feedback loss.

- An oscillator that can provide a very high quality sine wave output is the Wien-bridge oscillator. When the power is first turned on, the gain is very high and the amplifier produces an output of many frequencies. These frequencies are then fed back through the lead-lag network to the non-inverting input. Only the frequency that is in-phase, or positive, will appear at the output of the lead-lag network to be amplified again. A couple of loops through the circuit will provide a high quality sine wave output. Of course, the gain will decrease as the output at the correct frequency increases. The gain is controlled by the negative feedback loop.

- A Colpitts oscillator, with its two series capacitors, is susceptible to loading. Oscillators often use a buffer amplifier between their output and the load. A Colpitts oscillator is usually used as a fixed-frequency oscillator. Varying the capacitance of the two capacitors poses difficult mechanical problems, such as keeping the ratio of their capacitances equal.

- An oscillator whose resonant frequency can be varied and still retain the characteristics of the Colpitts is the Clapp oscillator. The action of the two series capacitors is to act as a voltage divider to control the amount of feedback. The addition of the capacitor in series with the inductor provides an easy method of varying the frequency of operation. Correct sizing of the capacitor will cause the frequency to be determined largely by this series capacitor. This capacitor can then be variable, to change the frequency easily.

- An easy way to recognize a Hartley oscillator is to observe the two-section inductor in the tank circuit. The value of the inductance in each section of the tank controls the amount of feedback. Look for either two inductors with the feedback tap between them, or a single inductor with a tap on it.

- An Armstrong oscillator can be recognized by the transformer-coupled feedback connection. This coil is sometimes called a tickler coil. The amount of feedback is controlled by the coupling between the inductor and the tickler coil.

- Crystal oscillators are very accurate. The crystal replaces the tank circuit. The difference is that the Q of the crystal can be very high. This makes the bandwidth very narrow. As a result, the oscillator is very accurate. The resonant frequency of a quartz crystal is susceptible to changes in temperature. Very accurate quartz crystals are often placed in a temperature-controlled oven to maintain an exact frequency of oscillation.

- The 555 timer is a very useful device for many types of timing circuits. As we found out, it is also used as a multivibrator. The multivibrator can be an astable type or a one-shot type. To quickly recognize an astable type, look for two external resistors and a capacitor. Another key is that the trigger input is not connected to a trigger circuit but is connected to the capacitor. A one-shot is recognized by having only one resistor and capacitor. The trigger is connected to an incoming signal.

CHAPTER 17 QUIZ

Student Name _____ Date _____

1. Oscillators operate on negative feedback.
   a. true
   b. false

2. The start-up gain of an oscillator must be greater than 1.
   a. true
   b. false

3. The feedback signal in a Hartley oscillator is derived from an inductive voltage in the tank circuit.
   a. true
   b. false

4. Crystal oscillators are inherently unstable.
   a. true
   b. false

5. The frequency in a VCO can be varied with a DC voltage.
   a. true
   b. false

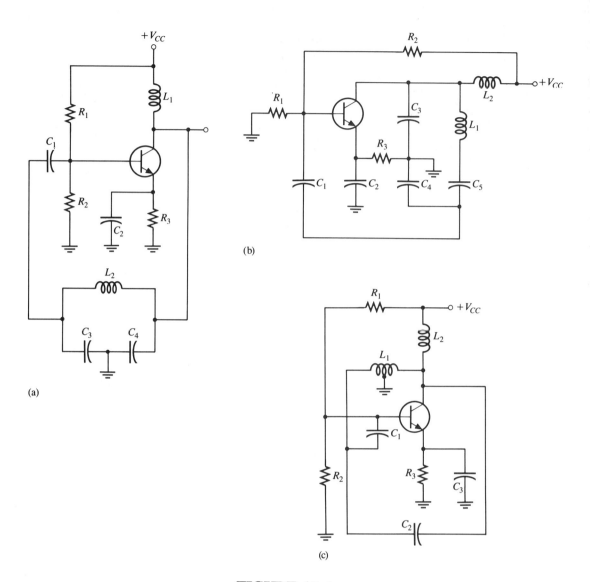

FIGURE 17-1

6. Refer to Figure 17-1(a). This circuit is known as
   a. a Clapp oscillator.
   b. an Armstrong oscillator.
   c. a Colpitts oscillator.
   d. a Hartley oscillator.

7. Refer to Figure 17-1(b). This circuit is known as
   a. a Clapp oscillator.
   b. an Armstrong oscillator.
   c. a Colpitts oscillator.
   d. a Hartley oscillator.

8. Refer to Figure 17-1(c). This circuit is known as
    a. a Clapp oscillator.
    b. an Armstrong oscillator.
    c. a Colpitts oscillator.
    d. a Hartley oscillator.

9. Refer to Figure 17-1(b). The resonant frequency is controlled by
    a. C_3 and L_1.
    b. C_2, C_4, C_5, and L_1.
    c. C_3, C_4, C_5, and L_1.
    d. C_3, C_4, C_5, and L_2.

10. Calculate the f_r of a lead-lag network if $R_1 = R_2 = 6.8$ kΩ, and $C_1 = C_2 = 0.05\mu F$.
    a. 468 Hz
    b. 4.68 kHz
    c. 46.8 kHz
    d. 468 kHz

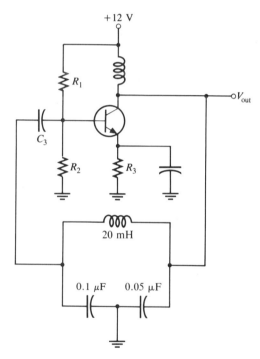

FIGURE 17-2

11. Refer to Figure 17-2. Calculate the resonant frequency.
    a. 1.126 kHz
    b. 6.17 kHz
    c. 23.9 MHz
    d. 14.1 MHz

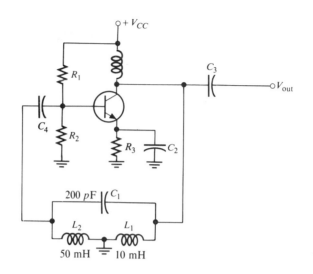

FIGURE 17-3

12. Refer to Figure 17-3. Determine the resonant frequency.
    a. 123.4 kHz
    b. 61.7 kHz
    c. 45.97 kHz
    d. 23.1 kHz

13. Refer to Figure 17-3. If C_1 increases in value, the resonant frequency will
    a. increase.
    b. decrease.
    c. remain the same.

14. A certain oscillator has a tap on the inductor in the tank circuit. This oscillator is probably
    a. a Colpitts oscillator.
    b. a Clapp oscillator.
    c. a crystal oscillator.
    d. a Hartley oscillator.

15. An op-amp integrator has a square-wave input. The output should be
    a. a sine wave.
    b. a triangle wave.
    c. a square wave.
    d. pure DC.

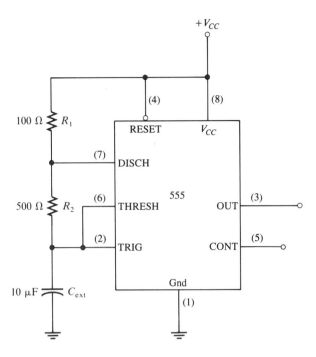

FIGURE 17-4

16. Refer to Figure 17-4. This circuit is
    a. sine-wave oscillator.
    b. a monostable multivibrator.
    c. an astable multivibrator.
    d. a VCO.

17. Refer to Figure 17-4. Determine the frequency of oscillation, if any.
    a. 131 Hz
    b. 262 Hz
    c. 2.62 kHz
    d. none.

18. Refer to Figure 17-4. What is the duty cycle?
    a. 16.3%
    b. 54.5%
    c. 86.9%
    d. 100%

19. Refer to Figure 17-4. If you desired to reduce the duty to less than 50%, the following circuit change would need to be made.
    a. Reduce the size of R_1.
    b. Reduce the size of R_2.
    c. Increase the size of R_1.
    d. Connect a diode in parallel with R_1.

20. A circuit that can change the frequency of oscillation with an application of a DC voltage is sometimes called
    a. a voltage-controlled oscillator.
    b. a crystal oscillator.
    c. a Hartley oscillator.
    d. an astable multivibrator.

REVIEW OF KEY POINTS IN CHAPTER 18

VOLTAGE REGULATORS

VOLTAGE REGULATION

- A **voltage regulation** circuit provides a constant DC output voltage with a limited input or load variation. The voltage regulator is an important part of a DC power supply.

- The need for voltage regulation occurs when the ac input source voltage changes or when the load current changes.

- When an ac input voltage varies, the regulation required is called **line regulation**. It is a calculated number expressed in percent (%).

- **Load regulation** is the change in output voltage as a result of a change in load current.

- The load regulation is expressed as a percentage and can be found by

$$\text{Load Regulation (\%)} = \frac{(V_{NL} - V_{FL})}{V_{FL}} \times 100$$

- A **voltage regulator** usually consists of a reference source, an error detector, a sampling element, and a control device.

- In many circuits, the reference source is a zener diode, the error detector is an op-amp, the sampling element is a voltage divider, and the control device is a power transistor.

- Many voltage regulator circuits incorporate **overcurrent protection circuits**.

BASIC SERIES REGULATORS

- A **series voltage regulator** uses a control element, reference voltage, error amplifier, and a sampling circuit.

- The series regulator gets its name from the series-connected control device. The control element is in series with the load.

- The sampling circuit is a voltage divider across the output.

- The output voltage can be found by $V_{OUT} \cong (1 + R_2/R_3)V_{REF}$, where R_2 and R_3 are the voltage divider resistors.

- The control device can be a power transistor mounted on a heat sink. The heat sink helps the transistor operate at a cooler temperature.

- An additional transistor, called a current limiter, is often used to provide **overload protection.**

BASIC SHUNT REGULATORS

- A **shunt regulator** has the control element connected in parallel with the load.

- Shunt regulators are not as widely used as the series type.

- This type of regulator does offer inherent short-circuit protection.

BASIC SWITCHING REGULATORS

- A more efficient type of voltage regulator is the **switching regulator**.

- The control element acts as a switch. Current is flowing when the element is on, and no current, when it is off. Since no power is used during the off cycle, efficiency is much greater than the series-type regulator, which dissipates power all the time.

- Switching regulators come in three basic configurations. The most common is the **step-down configuration**. There is also the **step-up configuration** and the **inverting type.**

- The regulator uses a voltage divider as a sensor circuit. The error voltage is amplified by an op-amp which causes a **pulse-width oscillator** to vary the pulse width supplied to the control element.

- The pulses are then filtered and supplied to the load. The regulation occurs as the average voltage supplied to the load is increased or decreased as sensed by the error amplifier.

IC VOLTAGE REGULATORS

- Integrated circuit manufacturers have built all of the above series regulators into a single package. These are called **three-terminal regulators**.

- The terminals are input, output, and ground.

- An entire family of both positive and negative regulators for fixed voltages is available.

- The 7800 and 7900 series are three terminal IC regulators with fixed positive and negative voltages respectively.

- These three-terminal regulators can have variable output voltages. This adds greater versatility to the power supply.

- The LM317 and LM337 series are three terminal variables with positive and negative output voltage respectively.

- A form of current limiting often built into three-terminal regulators is called **foldback current limiting**.

- This type of current limiting does not shut the regulator down completely when a short circuit occurs in the load. The current is reduced to a safe value which the regulator can handle without overheating.

- The 78S40 series is another type of IC regulator. This type is the switching voltage regulator.

TECHNICAL TIPS

- A voltage regulator is a circuit or device designed to keep an output DC voltage constant at a certain level regardless of changes in input voltage or load current changes. The load regulation is expressed as a percent; the lower the percentage, the better is the regulation. This means that, even if different loads are used, the output voltage stays the same.

- A series voltage regulator is often called a series-pass voltage regulator. The control element transistor is the pass transistor. These types of circuits work well. A common variation of this circuit is the addition of a potentiometer in the voltage divider sensor circuit. This allows the user to set the output voltage to a desired output voltage, within limits. Once adjusted, the circuit will then detect a voltage error and cause the pass transistor to conduct more or less heavily. This keeps the output voltage at a constant level.

- A switching regulator is very efficient, small, and lightweight. The switching transistor is either saturated (on) or cutoff (off). During the cutoff period, no power is dissipated. The saturation period consumes little transistor power, since V_{CE} is very small during saturation. This translates to high efficiency and accounts for its popularity.

  The frequency of operation of the pulse-width oscillator used in a switching regulator is quite high, around 25 kHz to 50 kHz. The filter components can be small, since the frequency is high. Typical values include an inductor of 22 mH and filter capacitors of around 100 μF. The advantages of excellent filtering action using an LC type of filter can be utilized because the components are very small.

- The three-terminal voltage regulator enjoys a very popular usage. These devices come in a variety of voltage outputs and in current ranges up to about 1.5 A. A power supply requiring more than this current can use a three-terminal regulator which in turn drives an external pass transistor. Loads of up to about 5 A can easily be regulated this way. Greater loads can be handled by paralleling several pass transistors. Much greater load currents can be regulated in this fashion.

- Once installed by the technician, these three-terminal regulators are almost impervious to faults. They have overcurrent protection built in; in case of overloads, they just shut down. When the overload has cleared, the regulators turn on again. Care must be exercised, however, in soldering these devices. The leads are heavy wire, and large amounts of heat are required to melt the solder and make a good bond. This excess heat can damage the device. Use a heat sink device clamped to the leads to conduct this heat away from the body of the regulator.

CHAPTER 18 QUIZ

Student Name _____ Date _____

1. A basic voltage regulator consists largely of a reference source, an error detector, and a control device.
   a. true
   b. false

2. Switching regulators are not very efficient.
   a. true
   b. false

3. IC voltage regulators are sometimes called three-terminal regulators.
   a. true
   b. false

4. Load regulation is the percentage change in line voltage for a change in load voltage.
   a. true
   b. false

5. A pass transistor is sometimes used in a series regulator.
   a. true
   b. false

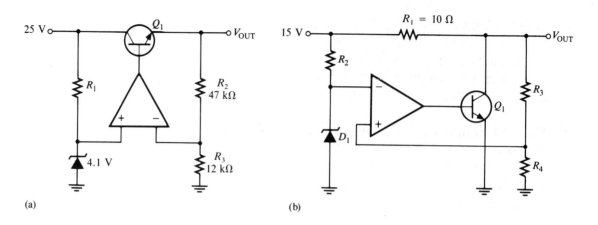
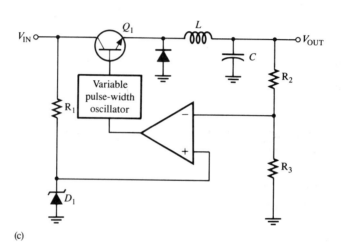

FIGURE 18-1

6. Refer to Figure 18-1(a). Determine the output voltage, V_{OUT}.
   a. 25 V
   b. 16.1 V
   c. 20.2 V
   d. 98 V

7. Refer to Figure 18-1(b). If the load is short-circuited, what would be the maximum current through R_1?
   a. 1 A
   b. 1.2 A
   c. 1.5 A
   d. 5 A

8. Refer to Figure 18-1(c). This circuit is called
   a. a series-pass voltage regulator.
   b. a shunt voltage regulator.
   c. a step-up switching regulator.
   d. a step-down switching regulator.

9. Refer to Figure 18-1(b). This circuit is known as
   a. a series-pass voltage regulator.
   b. a shunt voltage regulator.
   c. a step-up switching regulator.
   d. a step-down switching regulator.

10. Refer to Figure 18-1(a). Identify this circuit as
    a. a series-pass voltage regulator.
    b. a shunt voltage regulator.
    c. a step-up switching regulator.
    d. a step-down switching regulator.

11. Refer to Figure 18-1(c). This circuit operates at a _____ frequency, and its efficiency is _____.
    a. low, low
    b. low, high
    c. high, high
    d. high, low

12. Refer to Figure 18-1(c). The inductor and capacitor are used for
    a. amplifying the error signal.
    b. controlling the load current.
    c. turning on the pulse-width oscillator.
    d. filtering the DC pulse output.

13. Refer to Figure 18-1(b). The purpose for the diode D_1 is
    a. to supply a reference voltage.
    b. to amplify the error signal.
    c. to sense the error signal.
    d. to limit the input voltage to the circuit.

14. A voltage regulator has a no-load output of 18 V and a full load output of 17.3 V. The percent load regulation is
    a. 0.25%.
    b. 96.1%.
    c. 4.05%.
    d. 1.04%.

15. A voltage regulator with a no-load output DC voltage of 12 V is connected to a load with a resistance of 10 Ω. If the load resistance decreases to 7.5 Ω, the load voltage will decrease to 10.9 V. The load current will be _____, and the percent load regulation is _____.
    a. 1.45 A, 90.8%
    b. 1.45 A, 10.09%
    c. 1.6 A, 90.8%
    d. 1.6 A, 10.09%

16. Refer to Figure 18-1(b). If the output of the circuit were to be a short-circuit, what power rating would R_1 need to have?
    a. 2.25 W
    b. 5 W
    c. 10 W
    d. 25 W

17. Refer to Figure 18-1(a). If the zener diode had a rating of 1.7 V, the output voltage would be
    a. 6.66 V.
    b. 8.36 V.
    c. 9.1 V.
    d. 25 V.

18. Refer to Figure 18-1(a). If the BE junction of Q_1 opens, the output voltage V_{OUT} will
    a. increase.
    b. decrease.
    c. remain the same.
    d. increase to 18 V.

19. Refer to Figure 18-1(a). If a solder splash shorted the ends of R_1 to each other,
    a. the op-amp would fail.
    b. Q_1 would open.
    c. the output voltage would not change.
    d. the zener would fail.

20. Refer to Figure 18-1(c). If the output voltage tends to decrease due to an increase in load current, the transistor will conduct for _____ time each cycle.
    a. a longer
    b. a shorter
    c. the same

APPENDIX

ANSWERS TO CHAPTER QUIZZES

Chapter 1		Chapter 2		Chapter 3		Chapter 4		Chapter 5		Chapter 6	
1	A	1	B	1	B	1	A	1	A	1	A
2	A	2	A	2	A	2	B	2	B	2	B
3	B	3	B	3	B	3	A	3	A	3	A
4	B	4	A	4	A	4	B	4	B	4	B
5	B	5	B	5	A	5	B	5	B	5	A
6	C	6	D	6	C	6	C	6	C	6	C
7	C	7	D	7	B	7	D	7	D	7	B
8	D	8	C	8	D	8	C	8	B	8	D
9	B	9	A	9	C	9	B	9	C	9	C
10	A	10	C	10	A	10	C	10	D	10	B
11	B	11	D	11	B	11	B	11	B	11	A
12	B	12	A	12	E	12	A	12	D	12	D
13	D	13	C	13	B	13	D	13	B	13	A
14	C	14	B	14	D	14	B	14	D	14	C
15	A	15	D	15	C	15	D	15	B	15	D
16	C	16	A	16	A	16	A	16	D	16	C
17	B	17	C	17	A	17	B	17	C	17	B
18	C	18	A	18	C	18	A	18	A	18	C
19	B	19	B	19	B	19	B	19	A	19	D
20	C	20	C	20	D	20	C	20	C	20	C

Chapter 7		Chapter 8		Chapter 9		Chapter 10		Chapter 11		Chapter 12	
1	B	1	B	1	A	1	B	1	A	1	B
2	A	2	A	2	B	2	A	2	B	2	A
3	A	3	A	3	B	3	B	3	A	3	B
4	B	4	A	4	A	4	A	4	A	4	A
5	A	5	B	5	B	5	A	5	B	5	A
6	D	6	C	6	C	6	D	6	E	6	C
7	C	7	D	7	D	7	B	7	B	7	B
8	A	8	A	8	B	8	C	8	C	8	A
9	A	9	C	9	D	9	A	9	D	9	B
10	A	10	B	10	A	10	C	10	A	10	D
11	B	11	B	11	A	11	B	11	B	11	B
12	C	12	C	12	C	12	B	12	C	12	C
13	B	13	A	13	D	13	A	13	A	13	D
14	D	14	A	14	A	14	B	14	B	14	D
15	C	15	D	15	A	15	C	15	C	15	A
16	A	16	C	16	C	16	B	16	B	16	D
17	B	17	B	17	B	17	C	17	D	17	B
18	C	18	D	18	D	18	B	18	B	18	A
19	D	19	C	19	C	19	C	19	D	19	B
20	C	20	B	20	C	20	D	20	B	20	D

Chapter 13		Chapter 14		Chapter 15		Chapter 16		Chapter 17		Chapter 18	
1	A	1	A	1	A	1	B	1	B	1	A
2	A	2	B	2	B	2	A	2	A	2	B
3	B	3	A	3	B	3	A	3	A	3	A
4	B	4	B	4	A	4	A	4	B	4	B
5	A	5	A	5	A	5	A	5	A	5	A
6	C	6	D	6	B	6	D	6	C	6	C
7	B	7	C	7	B	7	C	7	A	7	C
8	D	8	D	8	B	8	B	8	D	8	D
9	D	9	C	9	B	9	A	9	C	9	B
10	B	10	B	10	A	10	C	10	A	10	A
11	B	11	A	11	A	11	A	11	B	11	C
12	B	12	D	12	C	12	A	12	C	12	D
13	B	13	C	13	B	13	D	13	B	13	A
14	D	14	B	14	D	14	A	14	D	14	C
15	C	15	C	15	A	15	A	15	B	15	B
16	B	16	B	16	C	16	A	16	C	16	D
17	A	17	C	17	B	17	B	17	A	17	B
18	B	18	A	18	A	18	B	18	B	18	B
19	C	19	B	19	D	19	D	19	D	19	D
20	B	20	C	20	C	20	B	20	A	20	A